AF595473

Applications on Ultrasonic Wave

Applications on Ultrasonic Wave

Editor

Jaesun Lee

MDPI • Basel • Beijing • Wuhan • Barcelona • Belgrade • Manchester • Tokyo • Cluj • Tianjin

Editor
Jaesun Lee
Changwon National University
Korea

Editorial Office
MDPI
St. Alban-Anlage 66
4052 Basel, Switzerland

This is a reprint of articles from the Special Issue published online in the open access journal *Applied Sciences* (ISSN 2076-3417) (available at: https://www.mdpi.com/journal/applsci/special_issues/Applications_Ultrasonic_Wave).

For citation purposes, cite each article independently as indicated on the article page online and as indicated below:

LastName, A.A.; LastName, B.B.; LastName, C.C. Article Title. *Journal Name* **Year**, *Volume Number*, Page Range.

ISBN 978-3-0365-1120-7 (Hbk)
ISBN 978-3-0365-1121-4 (PDF)

Contents

About the Editor

Jaesun Lee is a professor in the School of Mechanical Engineering and currently serves as the Director of Smart Manufacturing Engineering on the regional innovation platform and Vice Dean of Planning, Changwon National University, Korea. He obtained his PhD from Pusan National University, Korea, in the year 2015. He worked at Korea Railroad Research Institute (KRRI), Korea, and he joined the school of mechanical engineering of Changwon National University (CWNU), South Korea, as a tenure track faculty. His research includes ultrasonic NDE and structural health monitoring, long range guided wave inspection technology, theoretical wave scattering analysis, tomographic defect imaging, structural analysis and nonlinear ultrasonics material characterization. He had a global PhD fellowship from the Korean government in 2011 for 3 years. Moreover, he has won more than 11 best paper and presentation awards from the Korean Society for Precision Engineering, the Korean Society of Mechanical Engineers, the Korean Society for Nondestructive Testing and the Korean Society for Railway. He is currently leading the NDE research group of CWNU and several national funding projects, including the National Regional Leading Research Center.

Preface to "Applications on Ultrasonic Wave"

This Special Issue presents the latest advances in the field of applications on ultrasonic waves, including nondestructive methodology to evaluate the mechanical properties, damage states, and material condition of engineering structures. This Special Issue is composed of five papers covering new insights into inspection applications on road, pipeline and material characterization.

The advent of material characterization and non-destructive evaluation by ultrasonic wave applications is detailed in this book. Features related to propagation and the scattering of ultrasonic waves, such as wave velocities, dispersion, scattered amplitudes, and attenuation, show high sensitivity to material condition. Ultrasonic waves are also widely used in medicine to obtain images of internal body structures, such as muscles, tendons, blood vessels, joints, and organs. The mechanism is that ultrasound pulses are sent into the tissue using a probe, and the reflected signals are recorded and analyzed in order to build the desired images of internal structures.

Jaesun Lee
Editor

Article

Dynamic Non-Destructive Evaluation of Piezoelectric Materials to Verify on Accuracy of Transversely Isotropic Material Property Measured by Resonance Method

Yu-Hsi Huang [1,*], Chien-Yu Yen [1] and Tai-Rong Huang [2]

[1] Department of Mechanical Engineering, National Taiwan University, Taipei 10617, Taiwan; d08522007@ntu.edu.tw

[2] Taiwan Semiconductor Manufacturing Company, Hsinchu 30078, Taiwan; nilson444@yahoo.com.tw

* Correspondence: yuhsih@ntu.edu.tw; Tel.: +886-233662690

Received: 23 June 2020; Accepted: 21 July 2020; Published: 23 July 2020

Featured Application: This paper provides the measurement methods and corresponding calculating formulas for piezoelectric material constants. The completely orthotropic material property, i.e., dielectric, piezoelectric, and elastic constants, is obtained and verified by piezoelectric dynamic characteristics. By means of use in anisotropic material property, the mode shape and corresponding natural frequency were determined by finite element numerical calculation. The difference is rare in comparison with the results obtained by optically dynamic vibration measurement. The influence on self-heating phenomenon of soft and hard piezoelectric ceramics is also discussed in the study.

Abstract: In many engineering applications of piezoelectric materials, the design and prediction of the dynamic characteristics depends on the anisotropic electromechanical material property. Through collecting the complete formula in literature and listing all the prepared specimens, transversely isotropic material constants were obtained and verified by dynamic non-destructive evaluation in the paper. The IEEE (Institute of Electrical and Electronics Engineers) resonance method was applied to measure and calculate the orthotropic material constants for piezoelectric ceramics. Five specimens need to be prepared for the measurements using an impedance analyzer, in order to obtain the resonant and anti-resonant frequencies from the modes of thickness extension, length-extension, thickness-shear extension, length-thickness extension, and radial extension. The frequencies were substituted into the formulas guided on the IEEE standard to determine the elastic, dielectric, and piezoelectric constants. The dynamic characteristics of soft and hard piezoelectric ceramics in the results from the finite element method (FEM), which is analyzed from the anisotropic material constants of the resonance method, were verified with the mode shapes and natural frequencies found by experimental measurements. In self-heating, considered as operating on resonant frequencies of piezoelectric material, the resonant frequency and corresponding mode shape calculated by the material constants from resonance method in FEM are more accurate than the material property provided by the manufacturer and literature. When the wide-bandwidth frequency is needed to design the application of piezoelectric ceramics, this study completely provided the measurement method and dynamic verification for the anisotropic electromechanically material property.

Keywords: piezoelectric materials; transversely isotropic material; IEEE resonance method; resonant and anti-resonant frequencies; dynamic non-destructive evaluation

---

1. Introduction

Piezoelectric materials are smart materials which serve as crucial components in sensors or actuators. When the need arises for vibration control in lightweight flexible structures with complex shapes, piezoelectric sensors and actuators can be used to prevent adding to the overall mass of the structure and to achieve precision control. Piezoelectric materials have superior electromechanical properties, are lightweight and easy and inexpensive to produce, and have low power consumption [1–4]. For instance, the puncture systems used in cellular biology research have precise and fine vibration control [5]. In addition to applying the vibration characteristics of inverse piezoelectricity, the puncture systems also apply energy extraction systems with direct piezoelectric effects. Recently, vibration energy harvesting devices using piezoelectric materials have garnered the attention of researchers. These devices convert the vibration energy of mechanical structures into usable electrical energy, which can detect the inhibiting effects of vibration signals [6]. Energy harvesting systems based on lead zirconate titanate (PZT) can be embedded into vibrating objects to generate tens to hundreds of microwatts of energy for low-power integrated circuits (ICs). Thus, piezoelectric materials can be combined with low-power ICs to create continuously-working miniature wireless sensors, such as implantable biomedical sensors and wearable devices [7], wireless sensor networks [8], and smart buildings and special structures [9–11]. The vibration characteristics of piezoelectric materials can be analyzed in advance using finite element software for greater convenience in the optimization of piezoelectric effects. To obtain accurate analysis results, the boundary conditions of the model must be clearly defined, and the right piezoelectric material parameters must be used. The parameters provided with commercially-available piezoelectric materials are often inadequate for numerical calculations. Complete dynamic characteristics cannot be derived from the mechanical parameters of anisotropic materials, and thus, measuring effective and complete material parameters is key to analysis accuracy.

In this study, the material constants of the piezoelectric specimens were measured using the resonance method (RM). This method has been widely applied. In 1954, Mason et al. [12] used the resonance characteristics of an equivalent circuit simulating a piezoelectric crystal and employed static methods, quasi-static methods, dynamic methods, and hydrostatic methods to measure the elastic, piezoelectric, and dielectric constants of 45° Rochelle salt. Based on these measurement methods, they then developed RM to measure the material properties of piezoelectric crystals. In this method, the elastic and piezoelectric constants of the piezoelectric crystal are measured and the dielectric constants are inversely calculated from measured capacitance. Although the compliance S_{11}^{E} and S_{11}^{D}, electromechanical coupling coefficients k_{33} and k, and permittivity ε_{33}^{T} were obtained, the measurement was lack of guideline for prepared the standard specimens. In 1961, the IRE Standards on Piezoelectric Crystals [13] presented five specimens with different geometric shapes, aspect ratios, polarization orientations, and electrode surface orientations. The mode shapes corresponding to these five specimens were the thickness extension (TE) mode, length extension (LE) mode, thickness shear extension (TS) mode, length thickness extension (LTE) mode, and radial extension (RAD) mode. As polarized PZT has a hexagonal lattice (6 mm), it has 10 material constants, including five elastic constants, three piezoelectric constants, and two dielectric constants. With the TE, LE, TS, LTE, and RAD modes, the resonant and anti-resonant frequencies can be measured, and then the material constants of its various forms can be inversely calculated.

In 1987, the IEEE (Institute of Electrical and Electronics Engineers) Standards [14] referred to the RM in the IRE Standards and established a complete resonance measurement method with complete assumptions regarding constitutive equations (d-form, e-form, g-form, and h-form) and simplified specimen dimensions for piezoelectric materials and five specimens with different geometric shapes, aspect ratios, polarization orientations, and electrode surface orientations. Some formula listed in Refs [13,14] are more complicated, therefore the simplified form was proposed by subsequent researcher. In 1989, Wang et al. [15] applied the thickness vibration theory in RM to the vibrations in a piezoelectric plate in which the length and width are assumed to be much greater than the thickness that they approach infinity. They used α-quartz with two different cuts to measure the elastic, piezoelectric, and dielectric constants of α-quartz. The material constants of Y-cut quartz were studied in the paper; hence, the method is not provided for obtained the complete transversely piezoelectric isotropic material constants. In 1998, Cao et al. [16] conducted a detailed study on specimens with two types of shear mode shapes: thickness shear and length shear. They examined specimens with different aspect ratios and compared their measurements. The experiment results revealed that only the length-to-thickness ratio has a direct relationship and influences the values of the resonant and anti-resonant frequencies. A pure shear mode shape can be generated at l/20t, thereby reducing the errors in the resonant and anti-resonant frequency measurements.

We used an impedance analyzer to measure the resonant and anti-resonant frequencies of piezoelectric materials and substituted them into the equations listed in RM to obtain 10 independent material constants. Although the material parameters of piezoceramics are practically independent from frequency in the order of several MegaHertz [17–19], in the frequencies of the order of kilohertz—several tens of kilohertz—the dependence on frequency is very strong. It performs on the main resonances in longitudinal, transverse, shear, or radial modes. We then substituted the measured material constants into finite element numerical calculations to verify the in-plane and out-of-plane resonant frequencies and mode shapes of the piezoelectric materials. The experimental measurements of the out-of-plane resonant frequencies and mode shapes were obtained using electronic speckle pattern interferometry (ESPI), whereas the in-plane vibrations were measured using ESPI and an impedance analyzer. A comparison of the experiment results and the numerical calculations revealed fairly good accuracy in the material constants obtained using RM.

2. Theory Based on Resonance Method

From the linearly electro-thermo-elastic model [20], the constitutive equations are represented as

$$T_{ij} = c_{ijkl}S_{kl} - e_{ijk}E_k - \alpha_{ij}\theta \tag{1}$$

$$D_i = e_{ijk}S_{kl} + \varepsilon_{ij}^S E_j - \lambda_i\theta \tag{2}$$

$$\eta = \alpha_{kl}S_{kl} + \lambda_k E_k + C_\theta\theta \tag{3}$$

$$q_i = -K_{ij}\theta_{,j} \tag{4}$$

In these equations, T_{ij}, S_{kl}, E_k, D_i, q_i and θ are the components of stress, strain, electric field, electric displacement, heat flux and temperature difference with initial condition, respectively. The c_{ijkl}, e_{ijk}, α_{ij}, ε_{ij}^S, λ_i, C_θ and K_{ij} are the components of elastic, piezoelectric, thermal expansion, permittivity, pyroelectric coefficients, and the thermal conductivity tensor, respectively. If neglecting the thermo-elastic and thermo-electrical effects, the piezoelectric *e*-form is

$$\begin{cases} T_{ij} = c_{ijkl}S_{kl} - e_{ijk}E_k \\ D_i = e_{ijk}S_{kl} + \varepsilon_{ij}^S E_j \end{cases} \tag{5}$$

and the *d*-form constitutive equations are, respectively,

$$\begin{cases} S_{ij} = s_{ijkl}T_{kl} + d_{kij}E_k \\ D_i = d_{ikl}T_{kl} + \varepsilon_{ik}^T E_k \end{cases} \tag{6}$$

The linear piezoelectric constitutive equations for a piezoceramic material with crystal symmetry class C_{6mm} are presented in the components of elastic (c_{pq}^E), piezoelectric (e_{pq}), and dielectric coefficients (ε_{pq}^S) respectively.

The IEEE Standards [14] indicate that measuring the complete material constants of piezoelectric ceramic material requires piezoelectric specimens with five different mode shape characteristics (TE, LE, TS, LTE, and RAD), each with its own geometric shape, aspect ratio, polarization orientation, and electrode surface orientation. We used an impedance analyzer (model: HP-4194A) to measure their resonant and anti-resonant frequencies and inversely calculate the material constants of the piezoelectric materials. Based on Equations (7), (8), (12), (13), (19), (20), (22), and (23), the characteristics of the impedance measurements for TE, LE, TS, and LTE are as shown in Figure 1. Only one set of resonant and anti-resonant frequency values (f_r, f_a) needs to be substituted into Equations (25) and (26). In Equations (25) and (26) for RAD, the impedance measurement characteristics, as shown in Figure 1, have two sets of resonant and anti-resonant frequency values (f_1^r, f_1^a, f_2^r, f_2^a). Below, we obtain the material constants from the specimens of the five vibration modes according to the explanations in Ref [14].

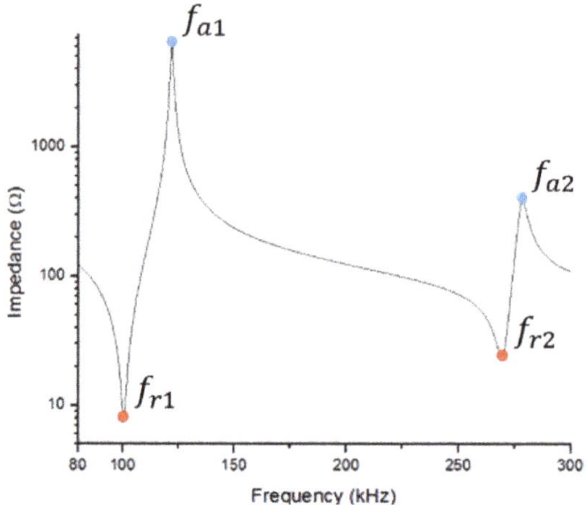

Figure 1. Schematics in resonant and anti-resonant frequencies obtained by impedance analyzer.

2.1. Thickness Extension (TE)

Table 1 presents the geometric shape and material constants of a thin square plate in which the directions of polarization and vibration are oriented along the depth (thickness). When the aspect ratios of the component satisfy $l > 10t$ and $w > 10t$, we use five specimens with dimensions $l = 46 \pm 0.1$ mm, $w = 46 \pm 0.1$ mm, and $t = 1.2$ mm, and the following equations can be used to calculate their material constants:

$$k_t^2 = \frac{\pi}{2}\frac{f_r}{f_a}\cot\left(\frac{\pi}{2}\frac{f_r}{f_a}\right) \tag{7}$$

$$c_{33}^D = 4\rho(tf_a)^2 \tag{8}$$

$$c_{33}^E = c_{33}^D\left(1 - k_t^2\right) \tag{9}$$

where k_t is the electromechanical coupling coefficient by thickness extension mode, $f_r$is resonant frequency, f_a is anti-resonant frequency, t is the thickness of specimen, and ρ is the density of specimen.

For the dielectric constant, the capacitance measured using an impedance analyzer at a specific frequency (1 kHz) can be substituted into the following equations, which will produce ε_{33}^T, the dielectric constant in constant stress, and ε_{33}^S, the dielectric constant in constant strain:

$$\varepsilon_{33}^T = \frac{C_l \cdot t}{A} \tag{10}$$

$$\varepsilon_{33}^S = \frac{C_h \cdot t}{A} \tag{11}$$

where A is the area of the electrode, C_l is the capacitance under 1kHz, and C_h is the capacitance at twice the anti-resonant frequency ($2f_a$).

Table 1. Specification on vibration modes of standard piezoelectric elements and their correspondent measured material constants [14].

Mode	Specification	Obtained Material Constant			
		k	d	s^E	ε^T
Thickness Extension (TE)	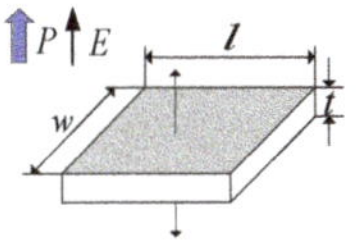 $l > 10t$, $w > 10t$	k_t	d_{33}	s_{33}^E	ε_{33}^T
Thickness Shear Extension (TS)	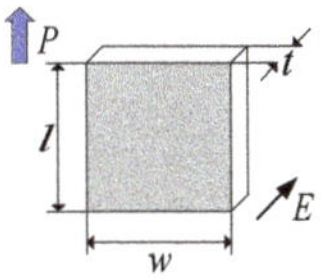 $l > 20t$, $w > 10t$	k_{15}	d_{15}	s_{44}^E	ε_{11}^T
Length Extension (LE)	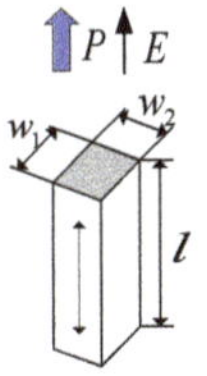 $l > 5w_1$, $w > 5w_2$	k_{33}	d_{33}	s_{33}^E	ε_{33}^T
Length Thickness Extension (LTE)	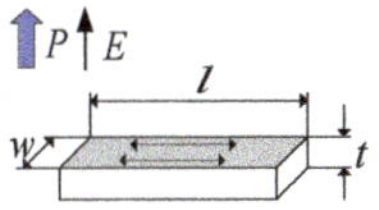 $l > 10t$, $l > 3w$, $3t > w$	k_{31}	d_{31}	s_{11}^E	ε_{33}^T
Radial Extension (RAD)	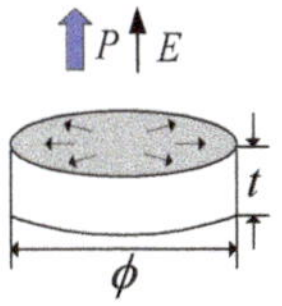 $\varnothing > 20t$	k_p	d_{31}	s_{11}^E	ε_{33}^T

E: Electric-field direction; **P**: Polarization direction.

2.2. *Thickness Shear Extension (TS)*

In this mode, we have a thin square plate in which the polarization is oriented along the length, and shear vibrations occur along the depth. In other words, the orientation of polarization is perpendicular to the direction of the electric field, causing the piezoelectric component to display shear vibrations. Table 1 presents the geometric shape and material constants. When the aspect ratios of the component satisfy $l > 20t$ and $w > 10t$, we use five specimens with dimensions $l = 9 \pm 0.05$ mm, $w = 8 \pm 0.05$ mm, and $t = 0.52$ mm, and the following equations can be used to calculate their material constants:

$$k_{15}^2 = \frac{\pi}{2}\frac{f_r}{f_a}\cot\left(\frac{\pi}{2}\frac{f_r}{f_a}\right) \tag{12}$$

$$c_{44}^D = 4\rho(tf_a)^2 \tag{13}$$

$$c_{44}^E = c_{44}^D\left(1 - k_{15}^2\right) \tag{14}$$

For PZT material, the following relations can be used to derive S_{44}^D, the elastic compliance constant with constant electric displacement, and S_{44}^E, the elastic compliance constant in a constant electric field:

$$S_{44}^D = 1/c_{44}^D \tag{15}$$

$$S_{44}^E = 1/c_{44}^E \tag{16}$$

where k_{15} is the electromechanical coupling coefficient of the thickness shear mode shape.

For the dielectric constant, the capacitance measured using an impedance analyzer at a specific frequency (1 kHz) can be substituted into the following equations, which will produce ε_{11}^T, the dielectric constant in constant stress, and ε_{11}^S, the dielectric constant in constant strain:

$$\varepsilon_{11}^T = \frac{C_L \cdot t}{A} \tag{17}$$

$$\varepsilon_{11}^S = \frac{C_H \cdot t}{A} \tag{18}$$

C_L is the capacitance measured at 1 kHz, and C_H is the capacitance measured at twice the anti-resonant frequency ($2f_a$).

2.3. Length Extension (LE)

In this mode, we have a thin square column in which the polarization and vibrations are oriented along the direction of the length. Table 1 presents the geometric shape and material constants. When the aspect ratios of the component satisfy $l > 5w_1$ and $w > 5w_2$, we use five specimens with dimensions $l = 10 \pm 0.1$ mm, $w_1 = 1.4 \pm 0.1$ mm, and $w_2 = 0.51$ mm, and the following equations can be used to calculate their material constants, where k_{33} denotes the electromechanical coupling coefficient of the longitudinal length mode shape, and l is the length of specimen.

$$k_{33}^2 = \frac{\pi}{2}\frac{f_r}{f_a}\cot\left(\frac{\pi}{2}\frac{f_r}{f_a}\right) \tag{19}$$

$$S_{33}^D = \frac{1}{4\rho(lf_a)^2} \tag{20}$$

$$S_{33}^E = \frac{S_{33}^D}{1 - k_{33}^2} \tag{21}$$

2.4. Length Thickness Extension (LTE)

In this mode, we have a thin square plate in which the polarization is oriented along the depth and lateral vibrations occur along the length. In other words, the vibrations occur along the length and perpendicular to the polarization orientation. Table 1 presents the geometric shape and material constants. When the aspect ratios of the component satisfy $l > 10t$, $l > 3w$, and $3t > w$, we use five specimens with dimensions $l = 31.8 \pm 0.1$ mm, $w = 3 \pm 0.1$ mm, and $t = 1.21$ mm, and the following equations can be used to calculate their material constants:

$$\frac{k_{31}^2}{k_{31}^2 - 1} = \frac{\pi}{2}\frac{f_a}{f_r}\cot\left(\frac{\pi}{2}\frac{f_a}{f_r}\right). \quad (22)$$

$$S_{11}^E = \frac{1}{4\rho(lf_r)^2} \quad (23)$$

$$S_{11}^D = S_{11}^E\left(1 - k_{31}^2\right) \quad (24)$$

where k_{31} denotes the electromechanical coupling coefficient of the thickness shear mode shape.

2.5. Radial Extension (RAD)

In this mode, we have a thin round plate in which the polarization and vibrations are both oriented along the depth. Table 1 presents the geometric shape and material constants. The aspect ratios of the component must satisfy $\varnothing > 20t$, and we use five specimens with dimensions $\varnothing = 19.9 \pm 0.05$ mm and $t = 0.81$ mm. In 1973, Meitzler, O'Bryan, and Tiersten [21] derived the constants of radial vibrations in round plates. The formula for k_p, the electromechanical coupling coefficient of the radial vibrations in a piezoelectric round plate is

$$\frac{k_P^2}{1 - k_P^2} = \frac{\left(1 - \sigma^E\right)J_1[\eta_1(1 + Hf/f_r)] - \eta_1(1 + \Delta f/f_r)J_0[\eta_1(1 + \Delta f/f_r)]}{(1 - \sigma^E)J_1[\eta_1(1 + \Delta f/f_r)]} \quad (25)$$

where k_p denotes the electromechanical coupling coefficient of the radial mode shape; $\Delta f = f_a - f_r$ is the difference between the anti-resonant frequency and the resonant frequency; J_0 is Bessel Function of First Kind and Zero Order; J_1 is Bessel Function of First Kind and First Order; σ^E is Poisson's Ratio; and η_1 is the first positive root from the function $\left(1 - \sigma^E\right)J_1(\eta) = \eta J_0(\eta)$. However, in 2005, Zhang, Alberta, and Eitel [22] modified the electromechanical coupling relation above using approximation into

$$k_P^2 = \frac{1}{P}\cdot\frac{f_a^2 - f_r^2}{f_a^2} \quad (26)$$

where

$$P = \frac{2\left(1 + \sigma^E\right)}{\left\{\eta_1^2 - \left[1 - \left(\sigma^E\right)^2\right]\right\}} \quad (27)$$

Based on RM by IEEE standard [14,15], several formula were simplified also by Zhang et al. [22] and we used them to calculate some parameters, i.e., ε_{33}^T ε_{33}^S, ε_{11}^T.

RM uses the vibration characteristics corresponding to the five types of piezoelectric components above to derive the elastic, piezoelectric, and dielectric constants. Furthermore, Poisson's ratio is defined as the ratio of lateral strain to longitudinal strain:

$$\sigma^E = -\frac{S_{12}^E}{S_{11}^E} \quad (28)$$

where the superscript E indicates values measured in a fixed electric field. The negative indicates that the two vibrations are opposite in direction, which means that when the longitudinal vibration is extending, the lateral vibration is contracting, and when longitudinal vibration is contracting, the lateral vibration is extending. Obtaining the Poisson's ratio generally requires the measurement of the fundamental frequency and one overtone frequency of the piezoelectric round plate. The formula used is $\left(1 - \sigma^E\right)J_1(\eta) = \eta J_0(\eta)$, the source and usage of which are explained in Ref [23,24]. Then, η_1 and η_2

are calculated, and we can look up the fundamental frequency $f_r = f_1^r$ and the overtone frequency f_2^r in the table provided in the Appendix to determine the Poisson's ratio σ^E, which is substituted into Equation (25) to derive k_p, the electromechanical coupling coefficient of the radial mode shape.

Although RM can be used to obtain the material constants of piezoelectric materials, there are still some constants that cannot be measured via experiment, and some conversions between constants are required to derive the constants that cannot be directly measured before all of the elastic, piezoelectric, and dielectric constants are obtained. Below is the solved equation including the piezoelectric strain constant:

$$d_{ij}^2 = k_{ij}^2 \varepsilon_{ii}^T s_{jj}^E \tag{29}$$

where $ij = 31, 33, 15$. Based on the equation above, if k_{ij}, ε_{ij}^T, and s_{ij}^E (i.e., the electromechanical coupling coefficient, the dielectric constant in constant stress, and the elastic compliance constant in a constant electric field) are known, then d_{31}, d_{33}, and d_{15}, (i.e., the piezoelectric strain constants) can be obtained. Next, we can use the following equations to obtain the other elastic and compliance constants:

$$s_{12}^E = -\sigma^E \cdot s_{11}^E \tag{30}$$

$$s_{12}^D = s_{12}^E - k_{31}^2 \cdot s_{11}^E \tag{31}$$

$$s_{66}^E = s_{66}^D = \frac{1}{c_{66}^E} = \frac{1}{c_{66}^D} = 2\left(s_{11}^E - s_{12}^E\right) \tag{32}$$

$$s_{13}^D = -\left\{\frac{s_{11}^D + s_{12}^D}{2} \cdot \left(s_{33}^D - \frac{1}{c_{33}^D}\right)\right\}^{0.5} \tag{33}$$

$$s_{13}^E = s_{13}^D + k_{31} \cdot k_{33}\left(s_{33}^E \cdot s_{11}^E\right)^{0.5} \tag{34}$$

$$c_{11}^E = \frac{s_{11}^E s_{33}^E - \left(s_{13}^E\right)^2}{\left(s_{11}^E - s_{12}^E\right)\left[s_{33}^E\left(s_{11}^E + s_{12}^E\right) - 2\left(s_{13}^E\right)^2\right]} \tag{35}$$

$$c_{12}^E = \frac{-s_{12}^E s_{33}^E + \left(s_{13}^E\right)^2}{\left(s_{11}^E - s_{12}^E\right)\left[s_{33}^E\left(s_{11}^E + s_{12}^E\right) - 2\left(s_{13}^E\right)^2\right]} \tag{36}$$

$$c_{13}^E = \frac{-s_{13}^E}{s_{33}^E\left(s_{11}^E + s_{12}^E\right) - 2\left(s_{13}^E\right)^2} \tag{37}$$

Using Equations (30) through (37), we can obtain all of the elastic and compliance constants. We can then use $\varepsilon_{ik}^T = \varepsilon_{ik}^S + d_{ip} e_{kp}$ to obtain the piezoelectric stress constant e_{ij}. The primary calculation formulas are as follows:

$$e_{31} = d_{31}\left(c_{11}^E + c_{12}^E\right) + d_{33} c_{13}^E \tag{38}$$

$$e_{33} = 2 d_{31} c_{12}^E + d_{33} c_{33}^E \tag{39}$$

$$e_{15} = d_{15} c_{44}^E \tag{40}$$

3. Specimens and Experimental Techniques

This study performed the experimental measurements and material constant calculations of two piezoelectric ceramic materials using RM. The in-plane and out-of-plane vibration characteristics were measured using instruments to verify the accuracy of the material constants obtained using finite element analysis. Below, we introduce the piezoelectric materials and instruments used in this study.

3.1. Prepation on Piezoelectric Ceramics

To establish complete piezoelectric ceramic material constants, we used RM to perform measurements of five specimens with specific geometric shapes, aspect ratios, polarization orientations, and electrode surface orientations. As shown in Table 1, the specific mode shapes include the TE, LE, TS, LTE, and RAD modes.

We procured piezoelectric ceramic specimens that fit the specifications of the TE and RAD modes in RM from American Piezo Ceramics (APC) International, Ltd. and the Fuji Ceramics Corporation. The piezoelectric specimens for the LE and LTE modes were cut to the required aspect ratios using ultrasonic cutting. Finally, for the TS mode, we ground our own specimen, removed the electrodes, and redistributed the electrodes in the direction normal to that of polarization so that the aspect ratio, polarization orientation, and the direction of the electric field created by the electrodes would meet specifications for TS mode testing.

3.2. Impedance Analysis

We derived material constants using the theoretical equations in RM. The resonant and anti-resonant frequencies substituted into the equations were obtained by measuring the mode shapes with specific structures, namely the TE, LE, TS, LTE, and RAD modes, using an impedance analyzer.

An impedance analyzer uses fixed voltages to excite vibrations in the target object and then measures the continuous changes in electrical impedance at various frequencies, as shown in Figure 1. The electrical impedance can be regarded as the total impedance applied by the resistors, inductors, and capacitors in the target object on the alternating current. Coupling characteristics are evident in the conversion between electrical and mechanical energy in piezoelectric materials. Thus, the electrical impedance of piezoelectric materials differs from that of components without piezoelectric properties. Generally, the results will display a resonant frequency ($f_{resonance}$, f_r) or anti-resonant frequency ($f_{anti-resonance}$, f_a) at certain resonant frequencies in frequency-impedance characteristic diagrams.

3.3. Electronic Speckle Pattern Interferometry (ESPI)

We input the material constants derived from measurements and calculations using RM and the material parameters provided by manufacturers or literature into finite element software to analyze vibration characteristics and obtain resonant frequency values and mode shapes. Next, we used the vibration characteristics obtained using optical non-destructive testing to verify the accuracy of RM.

Derived from holography, ESPI involves the casting of two coherent light beams on the target object and then the differences and changes in the optical path lengths are used to obtain the deformation data of the target object. The measurement method commonly used when ESPI is applied to vibrations is the time-averaging method. Within the exposure time of the CCD (charge-coupled device), images of the vibrating object are captured at different times, and a zero-order Bessel function is used to modulate the interference fringes. The brightest region in the interference images is the nodal line, where the displacement is zero, and the light and dark fringes show the contours of displacement. The resolution of the measurement system depends on the light source wavelength. We set up a sub-micron level ESPI system so as to obtain the full-field distributions of micro-vibration displacements in the interference patterns [25,26]. We then compared the ESPI measurement results of the low-frequency out-of-plane vibrations and the high-frequency in-plane vibrations in the piezoelectric specimens with the FEM analysis results derived from the piezoelectric material constant inputs.

3.3.1. Out-Of-Plane Vibration

As shown in Figure 2, a beam splitter is used to create two coherent laser beams. The one applied to the object surface is the object beam, whereas the other, known as the reference beam, is projected on a reference plate so that it scatters to create a speckled image. The two coaxial beams reflect and focus on the photosensitive plane of a CCD camera and create interference images. Using differencing calculations to process the images then swiftly produces interference fringe patterns on the computer screen, thereby providing real-time measurements during experiments.

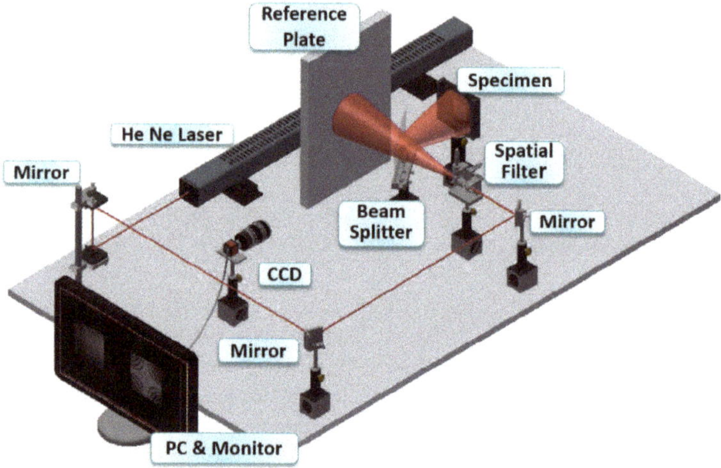

Figure 2. Out-of-plane setup in ESPI.

3.3.2. In-Plane Vibration

As shown in Figure 3, a beam splitter creates two laser beams which reach the target object from different sides but at identical angles of incidence with equal optical path lengths. Spatial filters then scatter the laser beams, and a CCD camera records the optimal interference images which undergo computer processing and produce vibration interference data.

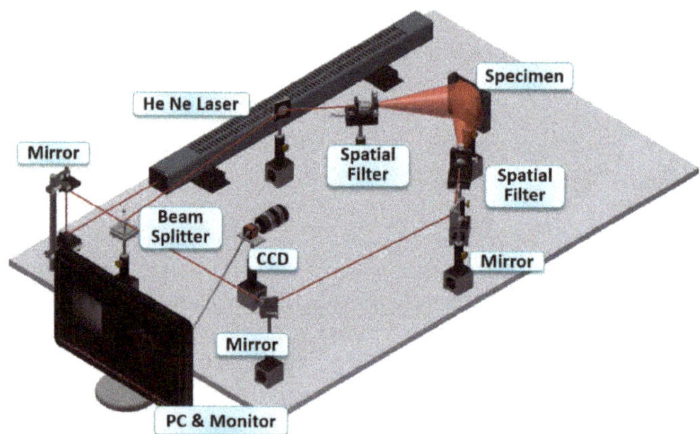

Figure 3. In-plane setup in electronic speckle pattern interferometry (ESPI).

4. Finite Element Method

We used the commercial software ABAQUS to perform finite element method (FEM) analysis calculations. For the material constants, we used the calculation results obtained using RM and the parameters provided by manufacturers and literature. It is noted that errors of about 6%-30% are produced on the natural frequencies of the first three modes, when the element used only one-layer thickness or the linear element is applied on the perturbation. To enhance the accuracy of the numerical calculation results and to confirm the convergence in calculation, we used high-order three-dimensional 20-node piezoelectric coupling elements with reduced integration (C3D20RE). The grid size was 1 mm in the length and width directions and was divided into five equal parts along the depth, resulting in 10,580 hexahedral elements comprising 50,243 nodes. We then compared the out-of-plane and in-plane mode shapes of the piezoelectric materials using FEM with experiment measurements to verify the accuracy of the material constants.

5. Results in Material Constants

The values required for RM were measured using an impedance analyzer and then converted. With the APC-855 material as an example, we took the average of the results from the five specimens. Table 2 presents the resonant and anti-resonant frequencies measured by the impedance analyzer in the TE mode, which were substituted into Equations (7) through (10) to derive K_t, C_{33}^E, and C_{33}^D. Table 3 presents the resonant and anti-resonant frequencies measured by the impedance analyzer in the TS mode, which were substituted into Equations (12) through (16) to derive K_{15}, C_{44}^E, and C_{44}^D. Table 4 shows the capacitance values measured at 1 kHz in the TE and TS modes, which were substituted into Equations (10), (11), (17), and (18) to derive ε_{33}^T and ε_{11}^T. Table 5 displays the resonant and anti-resonant frequencies measured by the impedance analyzer in the LE mode, which were substituted into Equations (19) through (21) to derive K_{33}, S_{33}^E, and S_{33}^D. Table 6 contains the resonant and anti-resonant frequencies measured by the impedance analyzer in the LTE mode, which were substituted into Equations (22) through (24) to derive K_{31}, S_{11}^E, and S_{11}^D. Table 7 shows the resonant and anti-resonant frequencies measured by the impedance analyzer in the RAD mode, which were substituted into Equations (25) through (27) to derive K_p and σ^E.

Table 2. Resonant and anti-resonant frequencies and the converted material constants in thickness extension (TE) mode.

TE Mode	f_r (kHz)	f_a (kHz)	k_t	$c_{33}^E (\times 10^{10} N/m^2)$	$c_{33}^D (\times 10^{10} N/m^2)$
Sample 1	1205	1875	0.7967	5.7156	15.6476
Sample 2	1162.5	1857.5	0.8093	5.2983	15.3568
Sample 3	1170	1875	0.8107	5.2763	15.3900
Sample 4	1190	1855	0.7976	5.5726	15.3155
Sample 5	1182.5	1850	0.8007	5.4972	15.2331
Avg. Value	–	–	0.8030	5.4720	15.3886

Table 3. Resonant and anti-resonant frequencies and the converted material constants in thickness shear extension (TS) mode.

TS Mode	f_r (kHz)	f_a (kHz)	k_{15}	$c_{44}^E (\times 10^{10} N/m^2)$	$c_{44}^D (\times 10^{10} N/m^2)$
Sample 1	1735	2100	0.6028	2.3078	3.6251
Sample 2	1720	2045	0.5804	2.2797	3.4377
Sample 3	1720	2055	0.5867	2.2765	3.4714
Sample 4	1760	2062.5	0.5607	2.3974	3.4968
Sample 5	1745	2092.5	0.5913	2.3407	3.5992
Avg. Value	–	–	0.5844	2.3204	3.5260

Table 4. Capacitance and dielectric constant measured at 1 kHz in TE and TS mode.

TE Mode	$C_l(nF)$	$\varepsilon_{33}^T(\times10^{-9}F/m)$	TS Mode	C_L (nF)	$\varepsilon_{11}^T(\times10^{-9}F/m)$
Sample 1	49.0810	28.1089	Sample 1	4.7493	30.9323
Sample 2	52.4846	29.9669	Sample 2	4.7647	31.0326
Sample 3	52.7355	29.9197	Sample 3	4.8435	31.4987
Sample 4	52.3802	29.8813	Sample 4	4.2233	27.5341
Sample 5	50.9098	29.1944	Sample 5	4.2036	27.3714
Avg. Value	–	29.4142	Avg. Value	–	29.6738

Table 5. Resonant and anti-resonant frequencies and the converted material constants in length extension (LE) mode.

LE Mode	f_r (kHz)	f_a (kHz)	k_{33}	$S_{33}^E(\times10^{-12}m^2/N)$	$S_{33}^D(\times10^{-12}m^2/N)$
Sample 1	132.75	170	0.6631	20.1927	11.3143
Sample 2	130	163.5	0.6453	20.9192	12.2074
Sample 3	134.5	169.25	0.6459	19.4678	11.3468
Sample 4	128.5	170.5	0.6946	21.7746	11.2705
Sample 5	133	172.5	0.6749	20.0189	10.9016
Avg. Value	–	–	0.6648	20.4746	11.4081

Table 6. Resonant and anti-resonant frequencies and the converted material constants in length thickness extension (LTE) mode.

LTE Mode	f_r (kHz)	f_a (kHz)	k_{31}	$S_{11}^E(\times10^{-12}m^2/N)$	$S_{11}^D(\times10^{-12}m^2/N)$
Sample 1	47.6	51.05	0.4019	14.2670	11.9626
Sample 2	47.75	50.9	0.3851	14.0978	12.0070
Sample 3	48.05	51.05	0.3756	14.1336	12.1402
Sample 4	47.6	50.6	0.3772	14.3118	12.2758
Sample 5	47.15	50.75	0.4114	14.4951	12.0416
Avg. Value	–	–	0.3902	14.2611	12.0854

Table 7. Resonant and anti-resonant frequencies and the converted material constants in radial extension (RAD) mode.

RAD Mode	f_1^r (kHz)	f_1^a (kHz)	f_2^r (kHz)	f_2^r/f_1^r	k_p	σ^E
Sample 1	100.35	121.8	269.2	2.6826	0.6365	0.2296
Sample 2	100.35	121.25	268.1	2.6716	0.6312	0.2439
Sample 3	100.35	121.8	269.2	2.6826	0.6365	0.2296
Sample 4	100.35	121.25	269.2	2.6826	0.6303	0.2296
Sample 5	100.9	119.6	266.45	2.6407	0.6100	0.2856
Avg. Value	–	–	–	–	0.6289	0.2437

Table 8 compares the material constants obtained using RM and those provided by the manufacturer for piezoelectric material APC-855 from APC International, Ltd., and Table 9 compares the material constants for piezoelectric material C-2 from the Fuji Ceramics Corporation. Some differences exist between the material constants obtained using RM in this study and those provided by the manufacturer or in literature for the two piezoelectric ceramic materials. Although the difference is not very large in piezoelectric *d*-form parameters, the finite element analysis produces calculating inaccuracy or it cannot obtain the results because of the missing terms in the manufacturer provided one. Due to transformation in calculation, the *e*-form parameters in the manufacturer provided one have a larger difference with those obtained by the resonance method in this study. We therefore also substituted these material constants into FEM to calculate the high-frequency in-plane low-frequency out-of-plane vibration characteristics and performed experimental measurements to determine the accuracy of applying these material constants to dynamic analysis.

Table 8. Material constant of APC-855 by resonance method, manufacturer, and Ref [27].

d Form	RM	Manufacturer Provided	*e* Form	RM	Ref [27]
Elastic compliance in constant electric field ($\times 10^{-12}$ m^2/N)			Elastic stiffness in constant electric field ($\times 10^{10}$ N/m^2)		
s^E_{11}	14.261	16.295	C^E_{11}	8.089	12.18
s^E_{12}	−3.475	–	C^E_{12}	2.451	8.021
s^E_{13}	−3.645		C^E_{13}	1.876	8.527
s^E_{33}	20.475	23.518	C^E_{33}	5.552	11.644
s^E_{44}	43.096	–	C^E_{44}	2.320	2.299
s^E_{66}	35.473	–	C^E_{66}	2.819	2.080
Piezoelectric strain constant ($\times 10^{-12}$ C/N)			Piezoelectric stress constant (C/m^2)		
d_{15}	661	720	e_{15}	15.338	14.26
d_{31}	−253	−276	e_{31}	−16.984	−10.48
d_{33}	516	630	e_{33}	19.156	16.60
Dielectric constant in constant stress ($\times 10^{-9}$ F/m)			Dielectric constant in constant strain ($\times 10^{-9}$ F/m)		
ε^T_{11}	29.674	–	ε^S_{11}	19.536	24.082
ε^T_{33}	29.414	29.218	ε^S_{33}	10.936	23.020

Table 9. Material constant of FUJI C-2 by resonance method and manufacturer.

d Form	RM	Manufacturer Provided	*e* Form	RM	Manufacturer Provided
Elastic compliance in constant electric field ($\times 10^{-12}$ m^2/N)			Elastic stiffness in constant electric field ($\times 10^{10}$ N/m^2)		
s^E_{11}	12.881	14.106	C^E_{11}	12.085	7.3
s^E_{12}	−3.834	−4.232	C^E_{12}	6.102	1.848
s^E_{13}	−5.355	−5.355 *	C^E_{13}	5.638	5.638 *
s^E_{33}	18.467	18.039	C^E_{33}	8.911	5.3
s^E_{44}	43.465	45.455	C^E_{44}	2.301	2.2
s^E_{66}	33.430	36.690	C^E_{66}	2.991	2.726
Piezoelectric strain constant ($\times 10^{-12}$ C/N)			Piezoelectric stress constant (C/m^2)		
d_{15}	573	692	e_{15}	13.183	15.224
d_{31}	−150	−158	e_{31}	−9.126	−9.126 *
d_{33}	322	367	e_{33}	11.779	11.779 *
Dielectric constant in constant stress ($\times 10^{-9}$ F/m)			Dielectric constant in constant strain ($\times 10^{-9}$ F/m)		
ε^T_{11}	14.832	17.442	ε^S_{11}	7.278	6.907
ε^T_{33}	12.641	12.927	ε^S_{33}	6.112	6.112 *

* Combined calculation results from RM experimental measurements and material parameters provided by manufacturer.

6. Discussion and Verification

We used ESPI to measure the out-of-plane and in-plane vibration characteristics of piezoelectric materials. The white regions in the images were the nodal lines, and the first lines next to them show the sub-micron vibration displacements [25,26]. The resonant frequencies measured by the impedance analyzer were also listed. Next, we applied the material constants obtained using the mathematical calculations in RM to finite element analysis software (ABAQUS) to derive the out-of-plane and in-plane resonant frequencies and mode shapes. The bold black lines in the mode shape results were the nodal lines, and we compared these numerical calculation results with experimental measurements to verify the reliability of the material constants.

6.1. APC-855

The vibration verification specimen used in this study was the thin square plate in the TE mode, in which the polarization and the vibrations are oriented along the depth. For the geometric dimensions, we used $l = 46$ mm, $w = 46$ mm, and $t = 1.2$ mm in the FEM calculations and conducted experimental verification.

6.1.1. Out-Of-Plane Vibration

As shown in Figure 4, nine mode shapes were measured in the out-of-plane vibrations, reaching around 10 kHz. "RM" indicates the results of inputting the parameters derived using RM into FEM calculations, and "Ref [27]" indicates the analysis results obtained in literature for APC-855. The simulation results of these two sets of material constants were both consistent with the mode shapes measured in the piezoelectric specimen experiments. However, the RM calculation results for the resonant frequencies were precise, the errors remaining within 6.472%. The errors in Ref [27], on the other hand, were all greater than 10% and increased as the frequency increased. This demonstrates that the material constants obtained using RM in this study have a certain degree of reliability.

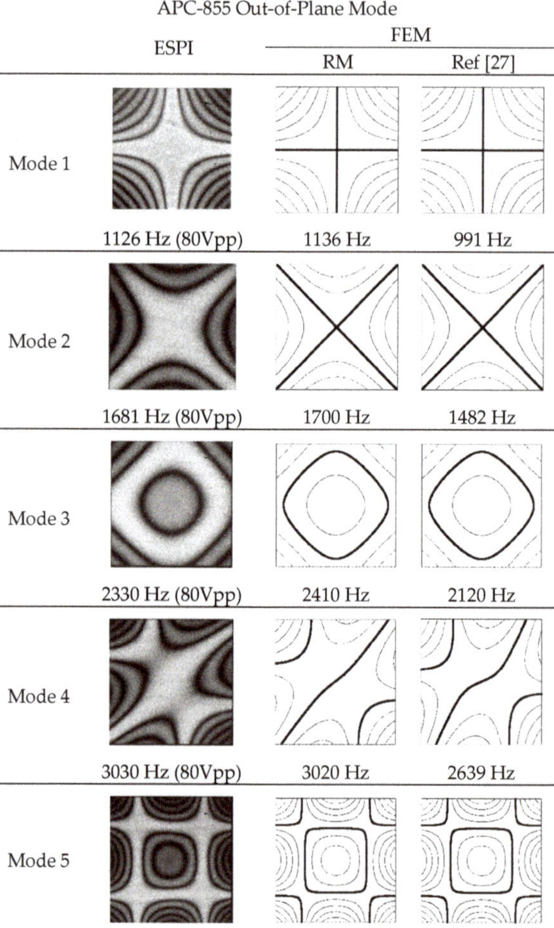

Figure 4. *Cont.*

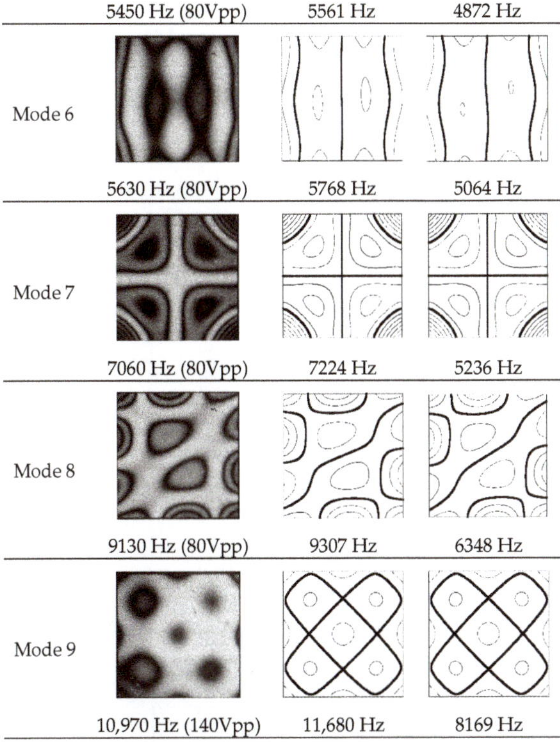

Figure 4. Out-of-plane vibration modes for APC-855.

6.1.2. In-Plane Vibration

As shown in Figure 5, five mode shapes could be obtained and the in-plane resonant frequency was in the high-frequency range from 20 kHz to 100 kHz. Mode shapes could be obtained via mode shape analysis of the material parameters derived in RM and Ref [27]. A comparison of the resonant frequency measurement results of impedance analysis and the ESPI frequency measurements revealed that they were almost identical. However, we found that the error between the RM analysis results and ESPI measurements increased as the frequency increased and exceeded 10%. In contrast, the analysis results in Ref [27] were relatively accurate. We speculate that this is because the impedance is smallest at resonant frequency vibrations; however, ESPI measurement of in-plane vibrations requires high voltages for excitation, which generates a self-heating effect and material property changes during resonance. We therefore speculate that the material constants derived in Ref [27] are better suited to the prediction of the in-plane vibrations in APC-855 following the self-heating effect. We explain this phenomenon in further detail in the following section.

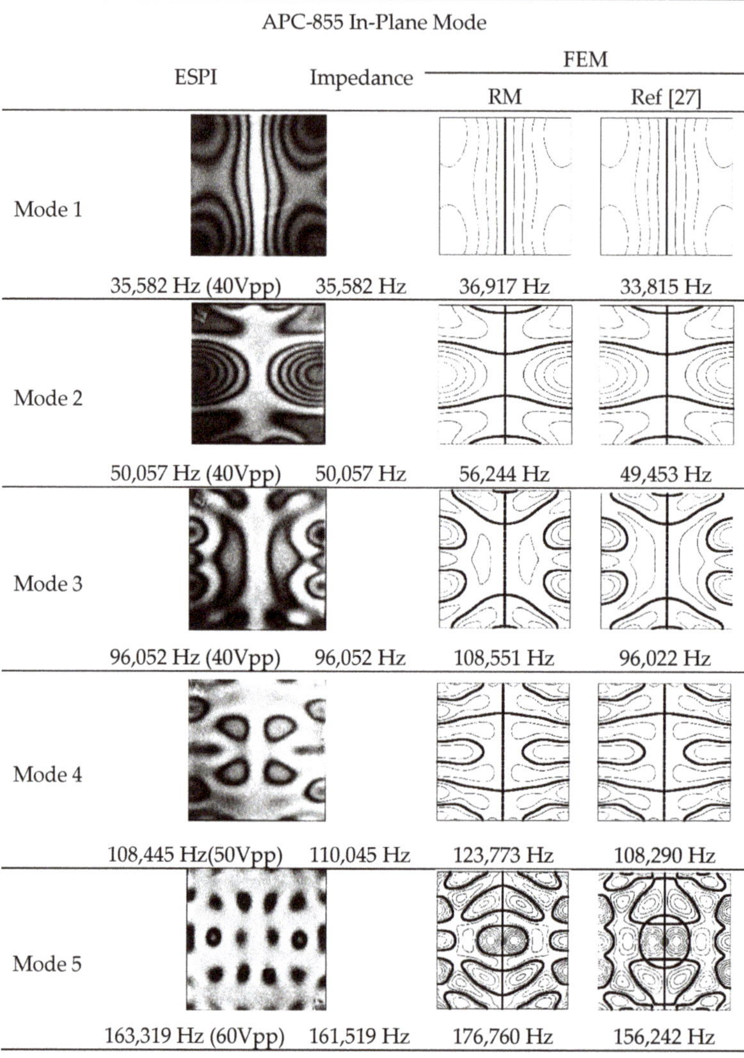

Figure 5. In-plane vibration modes for APC-855, after self heating in resonance.

6.1.3. Influence on Self-Heating in Resonance

For the out-of-plane modes, the piezoceramics was operated at lower frequencies and excited on almost one-hundred voltages. The piezoelectric material remained at almost the same temperature, which was measured by thermography at room temperature. However, we speculate that the in-plane parameters in Ref [27] are more accurate than the material constants obtained using RM in this study because APC-855 is a soft piezoelectric ceramic that is more prone to self-heating at high in-plane resonant frequencies. Thus, we used new specimens to conduct the in-plane vibration measurements. A comparison of Figures 5 and 6 show that the main difference lies in the duration of excitation. In Figure 6, the in-plane vibrations in the experiment were controlled so as to prevent self-heating caused by over-excitation [28,29]. The lower impedance occurs at the resonant frequency. The experimental measurement of in-plane vibrations requires higher excitation voltage, since ESPI measurement is sub-micrometer resolution. Under high-voltage, high-frequency, and low-impedance excitation conditions, high temperatures are often generated in piezoelectric components, and the material properties of piezoelectric components change at high temperatures, which may alter the capacitance and material rigidity of piezoelectric ceramic. The results in Tables 5 and 6 show that after self-heating, the errors between the resonant frequency measurements of APC-855 and the results of FEM calculations using the material constants obtained with RM become significant. When the voltage and excitation time are controlled under only several seconds to confirm slight temperature rising, the resonant frequencies from the experimental ESPI measurements and impedance analysis compared to FEM are almost identical. The RM results also become accurate, with errors less than 5.314%. This verifies that the parameters measured using RM can also be used to calculate in-plane vibration values with accuracy. The errors of the Ref [27] parameters input into FEM instead become higher than they were before the self-heating of the specimen, ranging from 7% to 11%. From Equations (1) through (4), the electro-thermal-elastic model illustrated that the physical parameters and material constants were influenced by a thermal effect. In the experimental results, it is concluded that material constants and dynamic characteristics of the piezoelectric material depend on temperature. Zhang et al. [22] also show the influence on dielectric loss, dielectric constant, shifting resonant frequency of piezoelectric material through the thermal effect in their results. Hence, the RM used in the measurement of piezoelectric material constants has to be considerate of the temperature dependency in the actual application. This study therefore demonstrated that the material constants obtained using RM can be used to accurately calculate the out-of-plane and in-plane vibration characteristics of piezoelectric materials. However, the impact of self-heating must be taken into account.

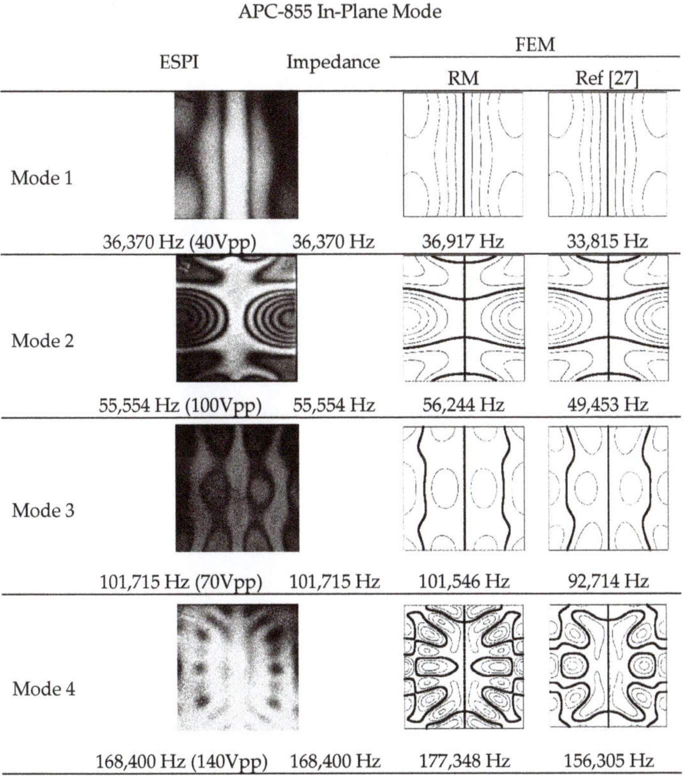

Figure 6. In-plane vibration mode for APC-855 before self heating in resonance.

6.2. FUJI C-2

In this paper, we discuss the accuracy of using RM to measure material constants of piezoelectric material APC-855, which is a soft ceramic in which self-heating is more severe during in-plane vibrations. We also examined the accuracy of using RM to measure material constants of piezoelectric material FUJI C-2, which is a hard ceramic. The dimensions of the test specimen were identical to those of the APC-855 specimen.

6.2.1. Out-of-Plane Vibration

As shown in Figure 7, ten mode shapes were obtained in the out-of-plane vibrations up to 12 kHz. We compared the experiment results and the FEM results. "Manufacturer provided" indicates the results of FEM calculations using the material constants provided by the manufacturer. However, not all of the parameters were available from the manufacturer, so we had to use parameters measured using RM to make up for the missing C_{13}^{E}, e_{31}, e_{33}, and ε_{33}^{S} in the FEM calculations of the vibration characteristics. Comparisons of the out-of-plane mode shape revealed high accuracy in all modes except Mode 5, which could not be derived using the parameters provided by the manufacturer. The RM errors were all less than 3.765%, thereby indicating that the RM results were all highly accurate. In contrast, only the errors of the results from the parameters provided by the manufacturer in Modes 1, 2, 3 and 6 were less than 10%. The errors in the remaining modes were greater than 10%. The manufacturers provided their piezoelectric material constants by measured in standard specimens; however, the quality control in turns of mass production influenced the difference of material properties. We can thus confirm that this study successfully used RM to obtain reliable material constants for piezoelectric ceramic materials.

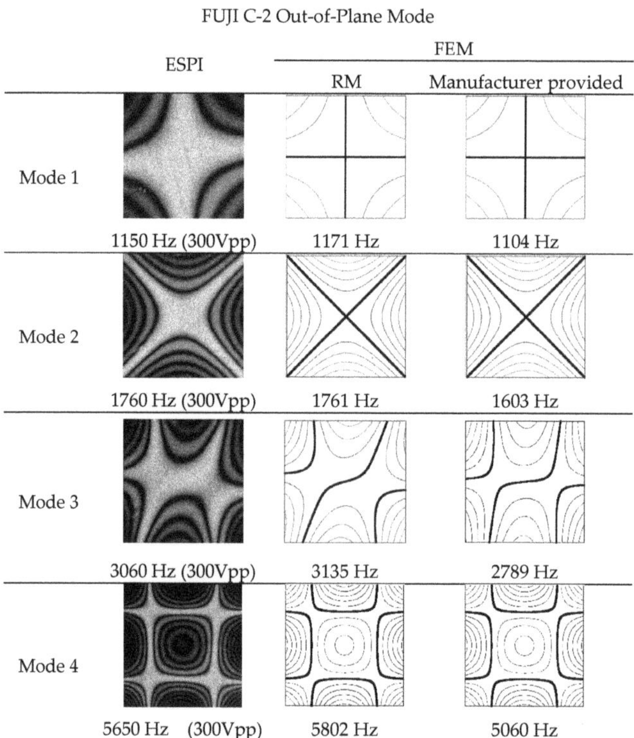

Figure 7. *Cont.*

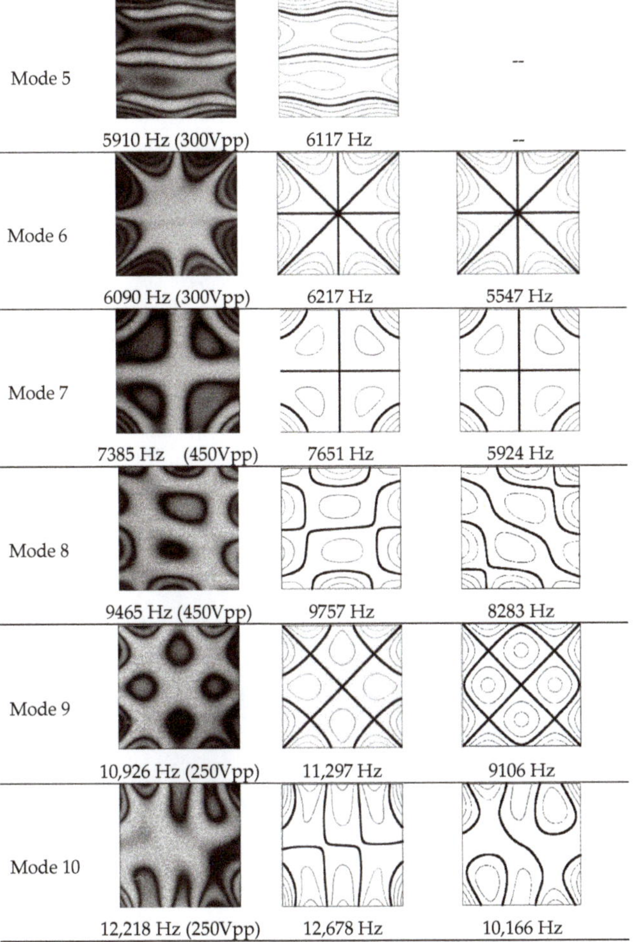

Figure 7. Out-of-plane vibration modes for FUJI C-2.

6.2.2. In-Plane Vibration

As shown in Figure 8, five mode shapes were obtained in the in-plane vibration experiments. The FEM results from RM material constants were all relatively accurate, whereas the parameters provided by the manufacturer could not produce any results at all. Because finite element analysis inputs the *e*-form material constants to calculate the results, it causes the transferred *e*-form material constants to have a larger difference with those that are manufacturer provided, as listed in Table 9. The impedance analysis results were also relatively consistent with the ESPI measurements, with only slight errors in Mode 5. The errors of the FEM calculations from the RM parameters were all within 4%, thereby indicating that the piezoelectric ceramic material constants obtained in this study were relatively reliable.

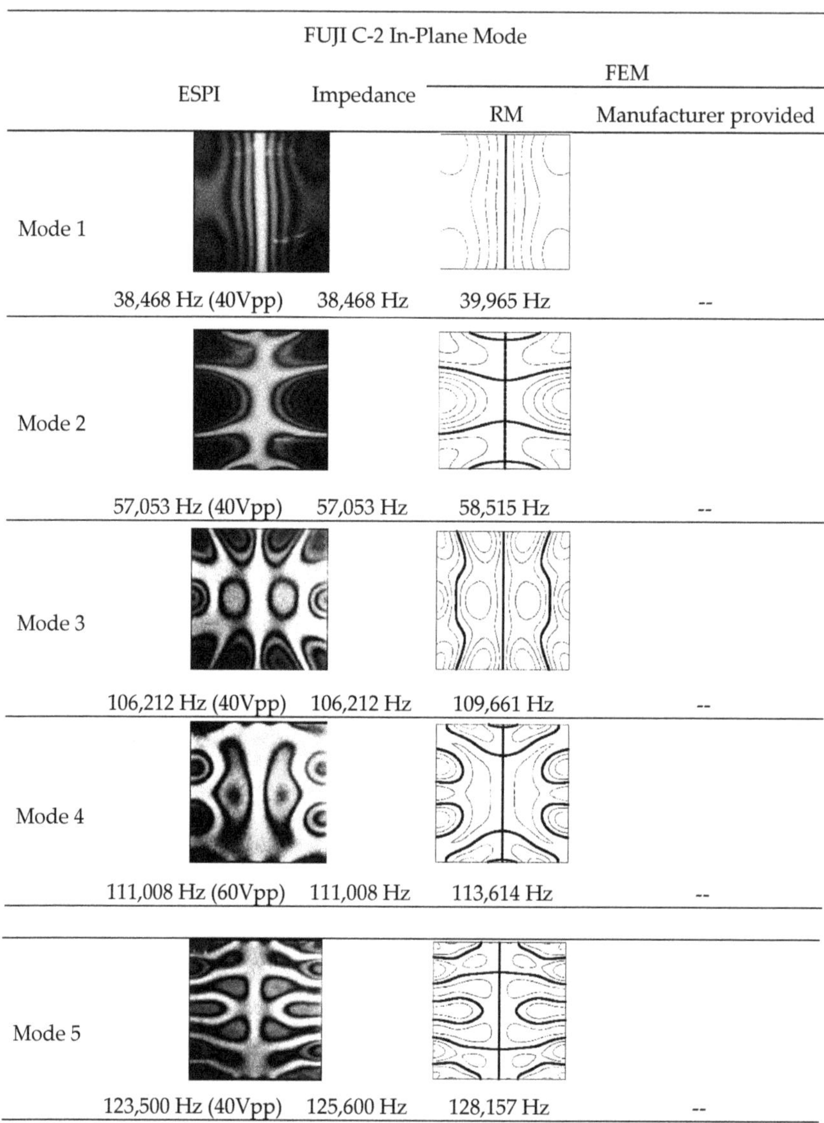

FUJI C-2 In-Plane Mode				
	ESPI	Impedance	FEM	
			RM	Manufacturer provided
Mode 1	38,468 Hz (40Vpp)	38,468 Hz	39,965 Hz	--
Mode 2	57,053 Hz (40Vpp)	57,053 Hz	58,515 Hz	--
Mode 3	106,212 Hz (40Vpp)	106,212 Hz	109,661 Hz	--
Mode 4	111,008 Hz (60Vpp)	111,008 Hz	113,614 Hz	--
Mode 5	123,500 Hz (40Vpp)	125,600 Hz	128,157 Hz	--

Figure 8. In-plane vibration modes for FUJI C-2.

6.3. Discussion on APC-855 and FUJI C-2

APC-855 and FUJI C-2 are soft and hard piezoelectric ceramics, respectively, and the applications of the piezoelectric effects of soft and hard piezoelectric ceramics differ. Soft piezoelectric ceramics present stronger direct piezoelectric effects and are thus more suitable for sensors, whereas hard piezoelectric ceramics show more distinct inverse piezoelectric effects and are thus more suitable for actuators. We examined the vibration measurements of these two materials via the drive voltage, and found that lower voltages can be applied to FUJI C-2 than to APC-855 while still producing ESPI measurement results at the sub-micro level. This verifies that actuators comprising hard ceramic piezoelectric materials may have stronger inverse piezoelectric effects. In contrast, soft ceramic piezoelectric materials require greater drive voltages to make the vibration fringes clear enough to observe, but the self-heating issue severely affects the resonant frequency measurement results. Once the temperature reaches half of the Curie Temperature and remains there, thermal deterioration takes place in the material and is accompanied by depolarization [30–33]. The vibration characteristics of the piezoelectric material are also affected, thereby affecting the accuracy of material constants derived using RM. Thus, this study successfully used dynamic characteristic measurement to verify the applicability of material constants obtained using RM.

7. Conclusions

Due to the anisotropic mechanical properties of piezoelectric materials, their material constants cannot be obtained using general material testing methods. Literature indicates that RM, an approach developed over years of research using structural and polarization characteristics, can effectively measure the anisotropic parameters of piezoelectric materials. This study contains complete details on the specimens and all of the equations that should be prepared, and we used dynamic experiments to verify the accuracy of the material constants that were obtained.

RM measurements require TE, LE, TS, LTE, and RAD mode specimens. Using impedance analysis, their resonant and anti-resonant frequencies were obtained and then substituted into the RM equations. We also used the conversional relationships in constitutive equations to obtain the complete elastic, piezoelectric, and dielectric coefficients of piezoelectric ceramic materials.

We used a non-destructive testing method for dynamic characteristic measurement, verifying FEM analysis results from anisotropic material constants obtained using RM from soft and hard piezoelectric ceramic materials. The errors between the out-of-plane and in-plane mode shapes, which is frequency-dependent on material constants, derived from the RM material constants and those measured in experiments, were very small. We also discussed the impact of the temperature changes caused by self-heating in soft and hard piezoelectric ceramics on the properties of the piezoelectric materials. Whether the piezoelectric ceramic is prone to self-heating affects the accuracy of RM material constants. The dynamic analysis of non-destructive testing revealed that soft ceramic is more prone to self-heating, which shifts the original resonant frequency. This indicates that the question of whether the properties of piezoelectric materials change due to self-heating must be noted, which will in turn affect the original operating frequency and voltage designs used in transducers.

Author Contributions: This study was initiated and designed by Y.-H.H., set-up experimental platform and conducted the experiments by C.-Y.Y. and T.-R.H. T.-R.H. were also responsible in numerical simulations. C.-Y.Y. and Y.-H.H. wrote the paper. All authors contributed in analyzing the simulation and experimental results. All authors have read and agreed to the published version of the manuscript.

Funding: This research was funded by Ministry of Science and Technology (Republic of China) under Grant MOST 107-2221-E-002 -193-MY3.

Acknowledgments: The authors gratefully acknowledge the financial support of this research by the Ministry of Science and Technology (Republic of China) under Grant MOST 107-2221-E-002 -193-MY3.

Conflicts of Interest: The authors declare no conflict of interest.

Appendix A

Table A1 is listed in the material property provided by APC International, Ltd. Table A2 is listed in the material property provided by Fuji Ceramics Corporation. Table A3 is listed in the Poisson's ratio, σ^E, correspondent to the ratio of the first two resonant frequencies.

Table A1. Material constants of APC-855 piezoceramic provided from APC International, Ltd.

Item	Unit	Symbol	APC-855
Electromechanical coupling coefficient	–	k_p	0.68
		k_{33}	0.76
		k_{31}	0.40
		k_{15}	0.66
Piezoelectric constant	$\times 10^{-12}$m/v	d_{33}	630
		d_{31}	−276
		d_{15}	720
Elastic constant	$\times 10^{10}$N/m^2	C^E_{33}	5.1
		C^E_{11}	5.9
Dielectric constant	@1 kHz	$\varepsilon^T_{33}/\varepsilon_0$	3300
Curie temperature	°C	T_e	200
Density	Kg/m^3	ρ	7600

Table A2. Material constants of APC-855 piezoceramic provided from Fuji Ceramics Corporation.

Item	Unit	Symbol	FUJI C-2
Electromechanical coupling coefficient	–	k_p	0.63
		k_{33}	0.76
		k_{31}	0.37
		k_{15}	0.77
		k_t	0.52
Piezoelectric constant	$\times 10^{-12}$m/v	d_{33}	367
		d_{31}	−158
		d_{15}	692
Elastic constant	$\times 10^{10}$N/m^2	C^E_{33}	5.3
		C^E_{11}	7.3
		C^E_{55}	2.2
Dielectric constant	@1 kHz	$\varepsilon^T_{33}/\varepsilon_0$	1460
		$\varepsilon^T_{11}/\varepsilon_0$	1970
Curie temperature	°C	T_c	300
Density	Kg/m^3	ρ	7600
Poisson's ratio	–	σ	0.3

Table A3. Poisson's ratio corresponds to the ratio of the first two resonant frequencies.

σ^E	η_1	η_2	f_2'/f_1'	σ^E	η_1	η_2	f_2'/f_1'
0.00	1.8412	5.3314	2.8956	0.26	2.0236	5.3817	2.6595
0.01	1.8489	5.3334	2.8846	0.27	2.0300	5.3836	2.6520
0.02	1.8565	5.3353	2.8738	0.28	2.0363	5.3855	2.6447
0.03	1.8641	5.3372	2.8632	0.29	2.0426	5.3874	2.6375
0.04	1.8716	5.3392	2.8527	0.30	2.0489	5.3894	2.6304
0.05	1.8790	5.3411	2.8425	0.31	2.0551	5.3913	2.6234
0.06	1.8864	5.3431	2.8324	0.32	2.0612	5.3932	2.6165
0.07	1.8937	5.3450	2.8225	0.33	2.0674	5.3951	2.6096
0.08	1.9010	5.3469	2.8127	0.34	2.0735	5.3970	2.6028
0.09	1.9082	5.3489	2.8031	0.35	2.0795	5.3989	2.5962
0.10	1.9154	5.3508	2.7936	0.36	2.0855	5.4008	2.5897
0.11	1.9225	5.3528	2.7843	0.37	2.0915	5.4027	2.5832
0.12	1.9296	5.3547	2.7750	0.38	2.0974	5.4046	2.5768
0.13	1.9366	5.3566	2.7660	0.39	2.1033	5.4066	2.5705
0.14	1.9436	5.3586	2.7570	0.40	2.1092	5.4085	2.5642
0.15	1.9505	5.3605	2.7482	0.41	2.1150	5.4104	2.5581

Table A3. *Cont.*

σ^E	η_1	η_2	f_2'/f_1'	σ^E	η_1	η_2	f_2'/f_1'
0.16	1.9574	5.3624	2.7396	0.42	2.1208	5.4123	2.5520
0.17	1.9642	5.3644	2.7311	0.43	2.1266	5.4142	2.5459
0.18	1.9710	5.3663	2.7226	0.44	2.1323	5.4161	2.5400
0.19	1.9777	5.3682	2.7144	0.45	2.1380	5.4180	2.5341
0.20	1.9844	5.3701	2.7062	0.46	2.1436	5.4199	2.5284
0.21	1.9911	5.3721	2.6981	0.47	2.1492	5.4218	2.5227
0.22	1.9977	5.3740	2.6901	0.48	2.1548	5.4237	2.5170
0.23	2.0042	5.3759	2.6823	0.49	2.1604	5.4255	2.5113
0.24	2.0107	5.3778	2.6746	0.50	2.1659	5.4274	2.5058
0.25	2.0172	5.3798	2.6670				

References

1. Nikkhoo, A. Investigating the behavior of smart thin beams with piezoelectric actuators under dynamic loads. *Mech. Syst. Signal Process.* **2014**, *45*, 513–530. [CrossRef]
2. Kant, M.; Parameswaran, A.P. Modeling of low frequency dynamics of a smart system and its state feedback based active control. *Mech. Syst. Signal Process.* **2018**, *99*, 774–789. [CrossRef]
3. Aridogan, U.; Basdogan, I.I. A review of active vibration and noise suppression of plate-like structures with piezoelectric transducers. *J. Intel. Mat. Syst. Str.* **2015**, *26*, 1455–1476. [CrossRef]
4. Huang, Q.Z.; Luo, J.; Li, H.; Wang, X.H. Analysis and implementation of a structural vibration control algorithm based on an IIR adaptive filter. *Smart Mater. Struct.* **2013**, *22*, 085008. [CrossRef]
5. Ru, C.G.; Pan, P.; Chen, R.H. The development of piezo-driven tools for cellular piercing. *Appl. Sci.* **2016**, *6*, 314. [CrossRef]
6. Feng, Q.; Fan, L.M.; Huo, L.S.; Song, G.B. Vibration Reduction of an existing glass window through a viscoelastic material-based retrofit. *Appl. Sci.* **2018**, *8*, 1061. [CrossRef]
7. Starner, T. Human powered wearable computing. *IBM Syst. J.* **1996**, *35*, 618–629. [CrossRef]
8. Roundy, S.; Wright, P.K.; Rabaey, J. A study of low level vibrations as a power source for wireless sensor nodes. *Comput. Commun.* **2003**, *26*, 1131–1144. [CrossRef]
9. Amirtharajah, R.; Chandrakasan, A.P. Self-powered signal processing using vibration-based power generation. *IEEE J. Solid State Circuits* **1998**, *33*, 687–695. [CrossRef]
10. Beeby, S.P.; Tudor, M.J.; White, N.M. Energy harvesting vibration sources for microsystems applications. *Meas. Sci. Technol.* **2006**, *17*, R175–R195. [CrossRef]
11. Ceponis, A.; Mažeika, D.; Bakanauskas, V. Trapezoidal cantilevers with irregular cross-sections for energy harvesting systems. *Appl. Sci.* **2017**, *7*, 134. [CrossRef]
12. Mason, W.P.; Jaffe, H. Methods for measuring piezoelectric, elastic, and dielectric coefficients of crystals and ceramics. *Proc. IRE* **1954**, *42*, 921–930. [CrossRef]
13. IRE standards on piezoelectric crystals: Measurements of piezoelectric ceramics. *Proc. IRE.* **1961**, *49*, 1161–1169. [CrossRef]
14. *IEEE Standard on Piezoelectricity*. ANSI/IEEE Std 176-1987. 1988. Available online: https://ieeexplore.ieee.org/document/26560 (accessed on 23 June 2020). [CrossRef]
15. Wang, J.F.; Chen, C.; Zhang, L.; Qin, Z.K. Determination of the dielectric, piezoelectric, and elastic constants of crystals in class 32. *Phys. Rev. B* **1989**, *39*, 12888–12890. [CrossRef] [PubMed]
16. Cao, W.W.; Zhu, S.; Jiang, B. Analysis of shear modes in a piezoelectric vibrator. *J. Appl. Phys.* **1998**, *83*, 4415–4420. [CrossRef]
17. Kitamura, T.; Kadota, M.; Kasanami, T.; Kushibiki, J.I.; Chubachi, N. Evaluation of piezoelectric ceramic substrates for ultrasonic bulk wave filters and resonators using pulse interference method. *Jpn. J. Appl. Phys.* **1994**, *33*, 3212–3216. [CrossRef]
18. Kuskibiki, J.; Akashi, N.; Sannomiya, T.; Chubachi, N.; Dunn, F. VHF/UHF range bioultrasonic spectroscopy system and method. *IEEE Trans. Ultrason. Ferroelectr. Freq. Control* **1995**, *42*, 1028–1039. [CrossRef]

19. Gudur, M.S.R.; Kumon, R.E.; Zhou, Y.; Deng, C.X. High-frequency rapid B-mode ultrasound imaging for real-time monitoring of lesion formation and gas body activity during high-intensity focused ultrasound ablation. *IEEE Trans. Ultrason. Ferroelectr. Freq. Control* **2012**, *59*, 1687–1699. [CrossRef]
20. Ahmad, S.N.; Upadhyay, C.S.; Venkatesan, C. Electro-thermo-elastic formulation for the analysis of smart structures. *Smart Mater. Struct.* **2006**, *15*, 401–416. [CrossRef]
21. Meitzler, A.H.; O'Bryan, H.M.; Tiersten, H.F. Definition and measurement of radial mode coupling factors in piezoelectric ceramic materials with large variations in poisson's ratio. *IEEE Trans. Sonics Ultrason.* **1973**, *20*, 233–239. [CrossRef]
22. Zhang, S.J.; Alberta, E.F.; Eitel, R.E.; Randall, C.A.; Shrout, T.R. Elastic, piezoelectric, and dielectric characterization of modified BiScO/sub 3/-PbTiO/sub 3/ ceramics. *IEEE Trans. Ultrason. Ferroelectr. Freq. Control* **2005**, *52*, 2131–2139. [CrossRef]
23. Erhart, J.; Pulpan, P.; Pustka, M. Piezoelectric ceramic resonators. In *Topics in Mining, Metallurgy and Materials Engineering*; Springer: Berlin/Heidelberg, Germany, 2017.
24. *Guide to Dynamic Measurements of Piezoelectric Ceramics with High Electromechanical Coupling*. IEC 60483:1976. 1976. Available online: https://webstore.iec.ch/publication/2229 (accessed on 23 June 2020).
25. Lin, Y.C.; Huang, Y.H.; Chu, K.W. Experimental and numerical investigation of resonance characteristics of novel pumping element driven by two piezoelectric bimorphs. *MDPI Appl. Sci.* **2019**, *9*, 1234. [CrossRef]
26. Huang, Y.H.; Ma, C.C. Experimental measurements and finite element analysis of the coupled vibrational characteristics of piezoelectric shells. *IEEE Trans. Ultrason. Ferroelectr. Freq. Control* **2012**, *59*, 785–798. [CrossRef] [PubMed]
27. Ma, C.C.; Chang, C.Y. Improving image-quality of interference fringes of out-of-plane vibration using temporal speckle pattern interferometry and standard deviation for piezoelectric plates. *IEEE Trans. Ultrason. Ferroelectr. Freq. Control* **2013**, *60*, 1412–1423.
28. Köhler, R.; RinderknechtInstitute, S. A phenomenological approach to temperature dependent piezo stack actuator modeling. *Sens. Actuators A Phys.* **2013**, *200*, 123–132. [CrossRef]
29. Quattrocchi, A.; Freni, F.; Montanini, R. Self-heat generation of embedded piezoceramic patches used forfabrication of smart materials. *Sens. Actuators A Phys.* **2018**, *280*, 513–520. [CrossRef]
30. Zhao, H.Y.; Hou, Y.D.; Zheng, M.P.; Yu, X.L.; Yan, X.D.; Li, L.; Zhu, M.K. Revealing the origin of thermal depolarization in piezoceramics by combined multiple in-situ techniques. *Mater. Lett.* **2019**, *236*, 633–636. [CrossRef]
31. Miclea, C.; Tanasoiu, T.; Miclea, C.F.; Amarande, L.; Cioangher, M.; Trupina, L.; Iuga, A.; Spanulescu, I.; Miclea, C.T.; David, C.; et al. Behavior of the main propeties of hard and soft type piezoceramics with temperature from 2 to 600 K. In Proceedings of the CAS 2010 Proceedings (International Semiconductor Conference), Sinaia, Romania, 11–13 October 2010.
32. Liao, Q.W.; Huang, P.F.; Ana, Z.; Li, D.; Huang, H.N.; Zhang, C.H. Origin of thermal depolarization in piezoelectric ceramics. *Scr. Mater.* **2016**, *115*, 14–18. [CrossRef]
33. Bobnar, V.; Kutnjak, Z.; Levstik, A. Temperature dependence of material constants of PLZT ceramics. *J. Eur. Ceram. Soc.* **1999**, *19*, 1281–1284. [CrossRef]

Article

An Application Study on Road Surface Monitoring Using DTW Based Image Processing and Ultrasonic Sensors

Sunil Kumar Sharma [1], Haidang Phan [2,3] and Jaesun Lee [4,*]

1 Extreme environment design and manufacturing innovation center, Changwon National University, Changwon 51140, Korea; sksharma@changwon.ac.kr
2 Institute of Theoretical and Applied Research, Duy Tan University, Hanoi 100000, Vietnam; phanhaidang2@duytan.edu.vn
3 Faculty of Civil Engineering, Duy Tan University, Danang 550000, Vietnam
4 School of Mechanical Engineering, Changwon National University, Changwon 51140, Korea
* Correspondence: jaesun@changwon.ac.kr

Received: 15 May 2020; Accepted: 28 June 2020; Published: 29 June 2020

Featured Application: Road surface monitoring for severe conditions.

Abstract: Road surface monitoring is an essential problem in providing smooth road infrastructure to commuters. This paper proposed an efficient road surface monitoring using an ultrasonic sensor and image processing technique. A novel cost-effective system, which includes ultrasonic sensors sensing with GPS for the detection of the road surface conditions, was designed and proposed. Dynamic time warping (DTW) technique was incorporated with ultrasonic sensors to improve the classification and accuracy of road surface detecting conditions. A new algorithm, HANUMAN, was proposed for automatic recognition and calculation of pothole and speed bumps. Manual inspection was performed and comparison was undertaken to validate the results. The proposed system showed better efficiency than the previous systems with a 95.50% detection rate for various road surface irregularities. The novel framework will not only identify the road irregularities, but also help in decreasing the number of accidents by alerting drivers.

Keywords: ultrasonic sensors; GPS; surface monitoring; image processing; dynamic time warping

---

1. Introduction

Treacherous road surface conditions are a significant problem for safe and comfortable transportation. The importance of road infrastructure for society can be compared with the importance of blood vessels for humans. For better road surface quality, it should be monitored continuously and repaired as necessary. The optimal distribution of resources for road repairs is possible by providing the availability of comprehensive and objective real-time data about the state of the roads. Developing countries with an escalating economy have an enormously extensive network of roads. Events within the past have shown that countries like the USA have invested about $68 billion a year on the maintenance of roads, bridges, and highways. These potholes also contribute to the degradation of suspension and fuel efficiency. In India, roads are the most common means of mobility; roads carry 90% of passengers and 65% of freight. Even with such a massive network, traveling is tiring in India as the roads are not sufficiently wide and the maintenance is inadequate. Regardless of where you are, driving is a stressful and potentially life-threatening affair. Over the last two decades, with an exponential rise in population, the number of vehicles has increased at a tremendous rate. The road infrastructure is not at par with the number of vehicles to support it, resulting in traffic congestion and accidents. As the ECONOMIC TIME's report of January 2018 suggests, at the very least, 400 people lose their lives every day in the country to road accidents due to potholes [1].

According to the report, "Road Accidents in India, 2016 [1] ", by the Ministry of Road Transport and Highways, there were 480,652 road accidents in 2016 as opposed to 501,423 in 2015. However, fatalities resulting from these accidents have risen by about 3.2%, which means that nearly 150,785 people were killed in 2016 compared to 46,133 in 2015 [1]. In India, most of the vehicles (unusually small hatchbacks) sold are either a 0-star rating or a 1-star rating in correspondence to the European standards, which implies that there could be fatal injuries even at moderate speeds. A considerable amount of research has been conducted on reducing traffic congestion and the number of ever-growing accidents. Vast sums of money are spent on repairing these potholes and bumps, however, small developing countries cannot afford to invest these amounts, which are a significant cause of road degradation. Irregularities on the road such as potholes caused by activities like erosion, weather, traffic, and some other factors may lead to accidents. Moreover, human-induced bumps such as speed breakers, which are made in the form of speed bumps, are used to control the speed of the vehicles, but they are not distributed at even distances, and the heights of the speed bumps have not been scientifically justified, which could also cause accidents. While said obstacles ought to be signaled according to specific road regulations, they are not always correctly labeled. One of the primary goals of this research was to detect road irregularities and decrease the number of accidents and fatalities. This approach is cost-efficient and can be applied worldwide. Generically, in modern-day field practices, the evaluation of the road irregularity data is done entirely via the raw data assessment [2]. Several road irregularity stats are used for the same. The existing technique can be categorically classified into three main types: imaging-based, sensor, and manual techniques.

There has been a substantial amount of research done in the past on the detection of potholes [3–5] using various techniques like reconstruction using laser [6], stereo vision [7,8], Light Detection and Raanging (LIDAR) [9], ultrasonic [10–13], etc. Forrest et al. [10] proposed a method of detecting road surface disruptions based on ultrasonic sensors in which the pothole detection rate of 62% was achieved on a paved road. Singh et al. [11] used a global positioning system receiver and ultrasonic for the identification of geographical location coordinates of the detected potholes and speed breaker with 59.23% efficiency. Madli et al. [12] worked on the automatic detection and notification of potholes and humps on roads to aid drivers using ultrasonic and a Global Positioning System (GPS). It was found that the system was 74% efficient and the data could be stored in the cloud. Shivaleelavathi et al. [13] also proposed an intelligent system using an ultrasonic sensor based on Raspberry Pi, GPS, and ultrasonic to take corrective measures. Devapriya et al. [14] proposed a real-time system that detected speed bumps by the analysis of photos of the road that were further enhanced using a Gaussian Filter (computer vision technique). This system had a correct rate ranging from 30% to 92%; however, failure was witnessed for faded speed bumps, and for a higher price, they had to be painted and maintained; this was a significant drawback of the system. Eriksson et al. [8] established the patrol system that comprised of a GPS and a 3-axis accelerometer for the detection of potholes. Likewise, Chen et al. [15] used the very same system and evaluated the power spectral density (PSD) for the detection of pavement irregularities using Fourier transformations. Inaccurate location of the irregularities was a shortcoming of the system. Nericell was a system developed by Mohan et al. [7] that comprised of amici, global system for mobile communication (GSM), accelerometer, and GPS sensors for road surface monitoring. The method yielded a False Negative Rate (FNR) of 37% and 41% for low speed and high-speed bumps, respectively. The use of smartphones, an adaptive fuzzy classifier, GPS, and inertial sensors for the detection of sudden driving events was presented by Arroyo et al. [16] and a precision performance of 0.87 and 0.91 was recorded. A fuzzy inference system was proposed by Aljaafreh et al. [17] for the detection of speed bumps using accelerometers embedded in smartphones. Nevertheless, the actual results of the tests were missing and accuracy only ranged from 70–80%. A multi-mobile phone system was developed by Astarita et al. [18], who also used accelerometers for the detection of bumps and potholes. A whopping 90% detection rate was obtained for bumps whereas the False Positive Rate (FPR) was about 35%.

Furthermore, the use of a smartphone accelerometer with six different algorithms to categorize road surface conditions was done by González et al. [19], who reported an Area Under Curve (AUC) ranging from 0.823 to 0.9445. Nature driven approaches such as genetic algorithm have also been explored: Yu and Yu [20] explored the use of genetic algorithms using images as a source of information to detect road irregularities. Moreover, abnormalities or distress data collection are increasingly being automated by using various imaging systems. However, analysis of the collected raw video clips for distress assessment is still predominantly being done manually, which is expensive, time-consuming, and slows down road maintenance management [21]. Pre-existing research suggests that three-dimensional image-based techniques are quite expensive; on the other hand, vibration sensor-based techniques are not as accurate and reliable [22] for road surface monitoring. The majority of the existing two-dimensional imaging methods have solely focused on either crack distress detection or pothole detection.

In this paper, an experimental analysis was conducted to detect potholes and speed bumps in quasi-real-time conditions using three different techniques, namely, an ultrasonic sensor with DWT, imaging-based systems, and manual inspection for validation from various roads of Noida city. In the first technique, which consists of an ultrasonic sensor, a GPS and Arduino connected to the Head Up Display (HUD) device are introduced. This system is a cost-effective method for a better, faster, and more reliable way to avoid accidents caused by potholes and raised humps. During the conduct of this experimental analysis, a collection of an exclusive data pattern of speed bumps and potholes was undertaken and directly fed to dynamic time warping (DTW) to find the likeness and similarity. A robust method for automated segmentation of frames with and without distress from the image processing technique is presented. These photo processing methods are responsible for the detection of potholes and speed bumps, which are supported by user-defined decision logic. The derivation of the decision logic is dependent on three characteristic visual properties of these irregularities (i.e., the standard deviation for calculating pixels intensities, circularity (CIRC) for calculating the shape, and average width for the dimension). An agile video separation algorithm called irregularity frame selection (IFS) was used to separate the visual frames with irregularities or not categorically. A new algorithm named the Hanuman algorithm took care of the second phase for automatically detecting and assessing potholes and speed bumps in one go. Finally, in the third technique, a manual visual inspection was undertaken to validate the above two methods.

2. Materials and Methods

The proposed threshold-based efficient road monitoring system detects the pothole and a speed break. Figure 1 shows the methodology overview of the information taken from the ultrasonic sensor. Collection and analysis of data were done using the primary server where the application of the DTW algorithm and noise filtering techniques was made. After the observation of the limitations of the existing systems, the present framework introduces the use of an ultrasonic sensor, DTW, IFS (algorithms), and filters for the detection of road irregularities. The ultrasonic sensor is fit in the moving direction of the car, as explained in Section 4. Information gathered with the help of sensors is sent to the central server for further analysis. Furthermore, noises, vibrations, and gravity components are eliminated using the filters. DTW helps to obtain the closeness score by analyzing the patterns for potholes and speed bumps. Ergo, road irregularity detection is undertaken with its GPS signature. Eventually, the final output is displayed on the mobile HUD for ease of the driver, making it even more ergonomic.

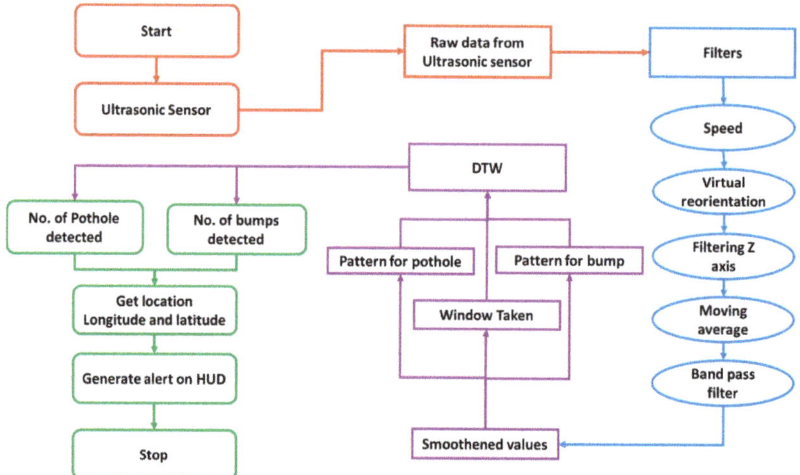

Figure 1. Flow chart of the proposed novel system includes ultrasonic sensor with dynamic time warping (DTW).

2.1. Hardware Setup

The main aim of the research was to create inexpensive hardware that runs on open software. Ergo, we used an Arduino Uno with an ultrasonic sensor (in Figures 2 and 3) and a GPS. Table 1 shows the hardware specification and Figure 4 shows the components used in the setup. The architecture of the proposed system consisted of four parts: an ultrasonic sensor, Arduino, a server module, and a heads-up display module. The ultrasonic sensor and Arduino are used to gather information about potholes and speed bumps and their geographical locations, and this information is sent to the server. The server module receives information from the microcontroller module, processes, and stores in the database.

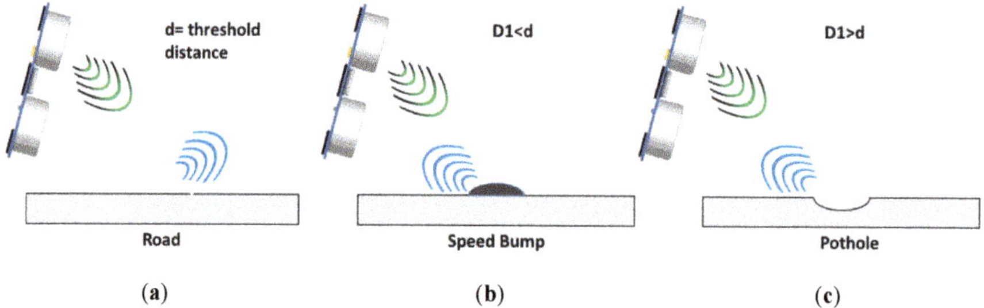

Figure 2. Working of ultrasonic sensor, (**a**) Plain road, (**b**) Speed bump, and (**c**) pothole.

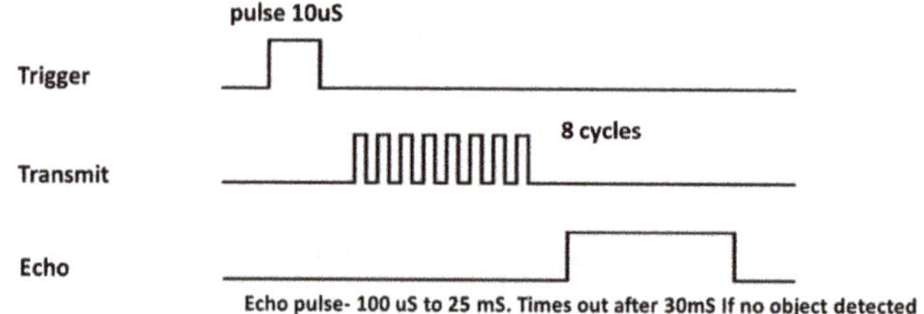

Figure 3. Timing response of the ultrasonic sensor.

Table 1. The hardware specification of the proposed framework.

Component	Hardware Description
Arduino	Microcontroller board based on ATmega328 using an open-source, prototyping platform, having 14 digital input/output pins (6 can be used as PWM outputs), 6 analog inputs, 16 MHz crystal oscillator, USB connection, power jack, ICSP header, and a reset button.
HY-SRF05	Narrow acceptance angle, 5-pin male header with 2.54 mm (0.1") spacing, 35 mA active, Range: approximately to 4 m (12- feet), Power requirements: +5 VDC; Communication: Dimensions: 22 × 46 × 16 mm (0.81 × 1.8 × 0.6 in), positive TTL pulse, Operating temperature range: 0 to + 70 °C (+32 to +158 °F). Figure 4a Arduino with HY-SRF05.
GPS Receiver	It is a satellite navigation system and is used to capture geographic location and time, irrespective of the weather conditions (Figure 4b).
HUD	A mobile device that used to display the information on the windshield as shown in Figure 4c.
Filters	Several filters like Speed, Separating Z-Axis, Moving Average, Band Pass, and Virtual Orientation are used.

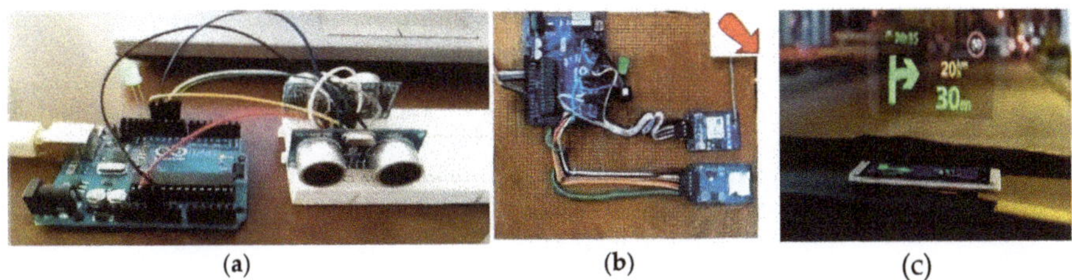
(a) (b) (c)

Figure 4. Components used in the road surface monitoring system: (**a**) Arduino with HY-SRF05; (**b**) GPS Receiver; (**c**) Head-up display module.

Ultrasonic sensors are utilized to determine the separation between the vehicle frame and the street surface numerically, and this value is passed to the microcontroller. Threshold distance is defined as the distance between the body of the vehicle and the smooth road surface, which depends upon the ground clearance of vehicles. If the value obtained by the ultrasonic sensor is more than the threshold distance, it recognizes it as a pothole, and if it is less, it is recognized as an unimportant irregularity, otherwise the street is smooth.

The microcontroller module (Arduino) is the main controller of the entire system and is responsible for incorporating the ultrasonic sensors and sending and receiving signals between the GPS receiver and the sensors. The microcontroller module is merely the processing power of the whole system and is utilized to accumulate data about potholes and their topographical areas, and these data are passed onto the server. The server module gets this data from the microcontroller and uploads its respective location signatures to the cloud server. This location will appear on the GPS system whenever any vehicle passes through or near it. The Arduino contains several filters to smoothen out the output from the ultrasonic sensor. The basic variations from the ultrasonic sensor are bolstered to the filters, where the information is smoothed and standardized. The filters are to filter the vibration due to some irregularities from the road surface apart from speed bumps and potholes. These vibrations are removed from the ultrasonic sensor information to lessen the impact of other outside powers and for the system to be able to differentiate between a constant/normal variation and an actual pothole or a bump.

2.2. Automated Assessment of Potholes and Speed Bumps Using Imaging Processing

The inception of the motivation for this research paper was induced directly or indirectly by our personal first-hand experiences with the Indian road conditions. The validation was for the determination of the efficiency of the proposed system. Initially, the pothole and bumps were detected with the help of using ultrasonic sensors paired with an Arduino attached with a heads-up display feeding the information to the GPS receiver. Irregularity frame selection is a recursive algorithm for searching all the vertices of a graph or tree data structure. This algorithm uses raw visual clips that are further put through several image processing methods; this is done using user-defined logics, which ensures the precise detection of irregularities. Segmentation of the irregular pixels in a visual frame is done by the algorithm using user defined decision logics that categorize those frames by the area covered by the irregular pixel. This segmentation is categorized into two parts: (1) Frames with irregularities, (2) frames without irregularities.

The Hanuman algorithm is used in the presented technique for computerized location and estimation of speed bumps and potholes in a single go from a grouping of video outlines. This algorithm is generated by the arrangement of two visual properties of bumps and potholes, which are distinguished from video photographs of Indian streets. This aggregate arrangement of recognized visual properties incorporates the accompanying, including the following:

Picture Texture: The picture surface of the interior of a pothole and speed bump is more different and distinguishing than the encompassing zone (i.e., irregularity less street surface region).

Shape Factor: The state of a pothole and speed bump is distinguishable. Their shape factor is estimated as far as circularity, which changes depending on their surface territory and border.

Measurement: The element of pothole and speed bump is expressed in terms of average width.

This particular algorithm fundamentally includes five segments:

- Image augmentation (Algorithm steps 3 and 4),
- Photo categorization (Algorithm step 5),
- Extraction of visual properties of articles (Algorithm steps 6 to 11),
- Recognition and classification of irregularities by decision logic (Algorithm step 12) and
- Quantification (Algorithm step 13).

The calculation produced for the robotized identification, estimation, and arrangement of casings with potholes and speed bumps without basic irregularity from a grouping of street video outlines is recorded underneath.

1. Categorized video frames are input;
2. The initial frame is selected;
3. A blue channel is selected for the conversion of the preset 24-bit depth format into 8-bit depth format;
4. Median filtering is used for photo augmentation;
5. Application of weighted mean based adaptive thresholding is made to covert processed photos into a binary image wherein black pixels signify objects of concentration;
6. Morphological erosion is used to add black pixels to link the slits in the binary photo;
7. Morphological dilation is used to eradicate secluded black pixels or their minor cluster;
8. Morphological erosion is used again to add black pixels to the binary photo;
9. Connected element classification and chain coding methods are used to quantify the number of substances or areas of notice and evaluate the area (A) and perimeter (P) of each of the object/substance;
10. Filter out all the objects/substances whose $A = 192\text{cm}^2$ in the primary photo;
11. Calculate standard deviation, circularity and average width of every remaining substance/object (critical objects, i.e., objects whose $A \geq 192\ \text{cm}^2$);
12. Categorize every substance/object into three kinds using heuristically derivative decision logic: Type (object) =
    (a) Pothole, if STD ≥ 12 & CIRC ≥ 0.14 & W ≥ 70 mm;
    (b) Speed Bump, if Convexity ≥ 5;

   (c) Unimportant Irregularity (UI), if otherwise;

13. Save the visual frame with the extracted and enumerated data in its respective category type folder;
14. Repeat steps 2 to 13 for all outstanding visual database;
15. Finish.

The procedure applied to extract the pothole information using the proposed algorithm is shown in Figure 5. Figure 5a is the original image with potholes; (b) is the binary image after median filtering (i.e., after random noise reduction in column (a) image), and adaptive thresholding (i.e., after object segmentation in column (b) image); (c) is binary image after erosion (i.e., after joining disparate black pixels in column (b) image); column (d) image after the Hanuman algorithm and (e) is the report of the pothole. Similarly, Figure 6 shows the speed bumps.

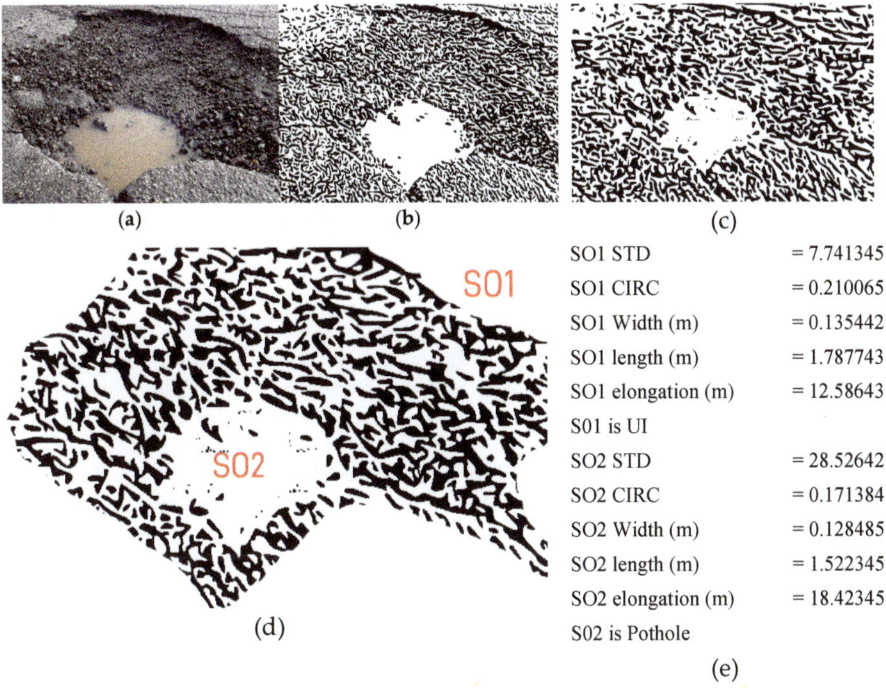

Figure 5. Procedure to categorize the image using the proposed method: (**a**) original image; (**b**) processed image; (**c**) further processed image; (**d**) image after Hanuman algorithm; (**e**) report of the Hanuman algorithm.

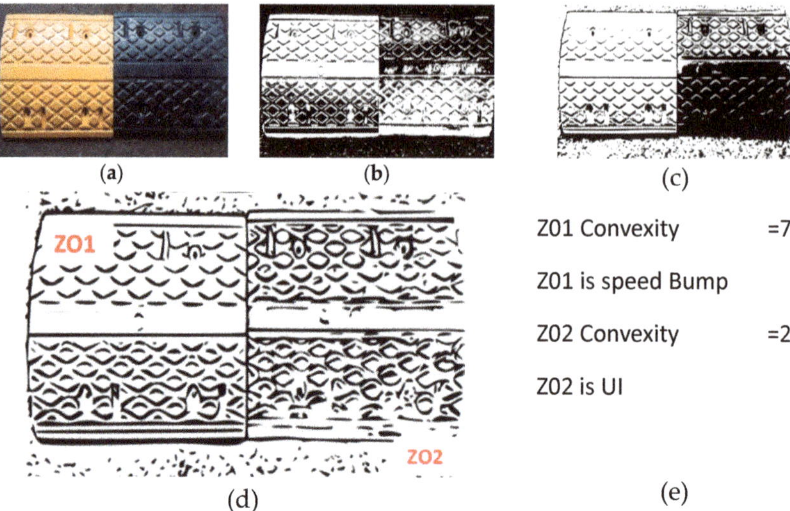

Figure 6. Procedure to categorize the image using the proposed method: (**a**) original image; (**b**) processed image; (**c**) further processed image; (**d**) image after Hanuman algorithm; (**e**) report of Hanuman algorithm.

3. Experimental Design and Location

To investigate the road surface, the framework used for the testing is shown in Table 2. The research was performed in the city of Noida, which comes under the National Capital Region of New Delhi. The city has an elevation of 200 m above sea level and a flat topographical region. Like most of the country, there are extremely contrasting roads in Noida, ranging from excellent roads to terrible roads. An expressway just outside of Noida (Figure 7a) and a slum road in Noida (Figure 7b) were considered for the experiment. Several test runs were performed, yielding different results. The experiment was conducted for three months covering around 150–170 km. For the evaluation of the performance of the framework, the algorithm presented in this research was applied in a Windows environment (Dell XPS 15Z with Intel i7 2600, 8GB DDR3 RAM on Windows 10 Pro) with the use of Visual Studio 2017 (VC ++ Redistributable) and Open CV3.3 Library. Hanuman, IFS, and DTW algorithms were used across the entire framework. For test runs, a Honda Amaze was used and image processing was done by a cost effective 4k resolution action camera. Table 2 shows the software and hardware specifications used in the experiment

Table 2. Software and hardware specifications used in the experimental setup

Technique Used	DTW + Ultrasonic Sensor + Image Processing
Location of Experiment	Noida City, UP, North India.
OS of the Data Collection System	Windows 10 Pro
Equipment used for Data Collection	Dell XPS 15z, HP Pavilion DV9000
Vehicle(s) used	Honda Amaze (2016)
Distance Covered (km)/Travel time (hours)	(150–170 km/4 h)
Detection	Location of potholes, speed bumps on the roads

(a)

(b)

Figure 7. Field test sites for road surface monitoring: (**a**) Yamuna expressway (from Noida to Agra, India); (**b**) potholes on the main roads (Khora, Noida)

4. Results and Discussion

The evaluation of the three different techniques is presented while traveling over the R1 (Yamuna expressway) and R2 (Khora Village) roads of Noida city and was also compared with other existing techniques.

4.1. Proposed Novel Framework

The ultrasonic sensor detects the pothole and speed bumps; these data are saved on the server and used to determine the separation amid the vehicle body and the street surface numerically, and this value is passed by the Arduino. Threshold distance is defined as the distance between the body of the vehicle and the smooth road surface and depends upon on the ground clearance of vehicles. If the value obtained by the ultrasonic sensor is more than the threshold distance, it will recognize it as a pothole, and if it is less, it is a mound, otherwise the street is smooth. The proto test was conducted using the design described in the Methodology Section above, and the results for R1 and R2 are shown in Tables 3 and 4.

Table 3. Information of pothole (P) and bumps (B) collected during real-time testing using an ultrasonic sensor with dynamic time warping (DTW) for the R1 test site.

Scenario No.	Obstacle Type	Depth/Height (cm)	Latitude (°N)	Longitudinal (°E)
1	B	3.0	28.4215	77.5261
2	B	3.0	28.4216	77.5261
3	P	2.6	28.4215	77.5265
4	B	5.2	28.4214	77.5266
5	B	4.8	28.4214	77.5268

Table 4. Information of pothole (P) and bumps (B) collected during real-time testing using ultrasonic sensor with DTW for R2 test site

Scenario No.	Obstacle Type	Depth/Height (cm)	Latitude (°N)	Longitudinal (°E)
1	P	9.27	28.6186	77.5544
2	B	2.20	28.6243	77.5661
3	B	8.63	28.6543	77.5667
4	P	13.12	28.6722	77.5643
5	P	8.20	28.6932	77.5634
6	P	8.19	28.7423	77.5612
7	B	3.20	28.7865	77.5755
8	P	1.77	28.7933	77.5765
9	B	3.72	28.8654	77.5659
10	P	12.16	28.9653	77.5760

As can be clearly seen, the uneven roads of the rural areas showed many potholes and bumps with some specific dimensions. On the other hand, the motorway road only showed irregularities up to 6.2 cm, which were filtered by filters such as Speed, Virtual Re-orientation, Filtering Z-pivot, SMA, and Band-Pass on the basis of the threshold limit being 8 cm. Figure 8 shows the experimental comparison in the z-axis displacement for R1 and R2.

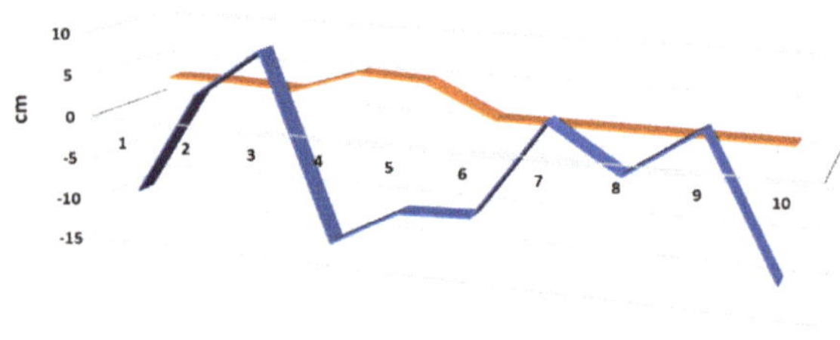

Figure 8. Experimental comparison of R1 (Yamuna expressway, urban site) and R2 (potholes on the main road (Khora, Noida, rural site)

4.2. Image Processing Result

Identification of the unique visuals of irregularities like illumination, size, and shape are made using image processing techniques. These accurately identify and help classify those irregularities. The irregularity frame selection algorithm is used as a quick video separation of the real-life visual clips captured of the Indian National Highways at several locations, which are divided into two frame categories (frames with irregularities and frames without irregularities). The Hanuman algorithm is responsible for the evaluation of the database of frames with irregularities, which is an automatic technique for the recognition and analysis of potholes and speedbumps. The proto test was conducted using the design described in the image processing methodology and the results for R1 and R2 are shown in Tables 5 and 6.

Table 5. Information of pothole (P) and bumps (B) collected during real-time testing using the image processing method for the R1 test site.

Scenario No.	Obstacle Type	Depth/Height (cm)	Latitude (°N)	Longitudinal (°E)
1	B	3.14	27.1937	76.5725
2	B	3.11	27.5405	76.4872
3	P	2.57	28.0691	76.4799
4	B	5.40	28.8080	76.6505
5	B	4.60	27.9979	77.1004

Table 6. Information of pothole (P) and bumps (B) collected during real-time testing using the image processing method for the R2 test site.

Scenario No.	Obstacle Type	Depth/Height (cm)	Latitude (°N)	Longitudinal (°E)
1	P	8.8695	27.4166	76.2903
2	B	2.1318	27.4507	75.7666
3	B	8.3590	27.6514	77.3884
4	P	12.9271	28.3281	76.5420
5	P	8.2082	27.7750	74.2999
6	P	7.8984	28.2824	75.7508
7	B	3.0682	28.0668	74.5035
8	P	1.7144	28.0620	75.3345
9	B	3.6010	27.9302	75.0760
10	P	11.6006	28.2817	75.0858

4.3. Manual Inspection

Manual inspection was performed using a band meter, which is used to measure the depth of an object with precision and accuracy. Figure 9 shows the manual inspection at R1 and R2, respectively. The proto test was conducted using manual inspection, and the results for R1 and R2 are shown in Tables 7 and 8. Generally, manual inspection is performed using some technical equipment, but in our case, we performed a very traditional manual inspection. Different scales were used to measure the different potholes and speed bumps for the validation.

(a)

(b)

Figure 9. Manual inspection on a rough patch in the middle of R1 (**a**) and R2 (**b**).

Table 7. Information of pothole (P) and bumps (B) collected by manual inspection for the R1 test site.

Scenario No.	Obstacle Type	Depth/Height (cm)	Latitude (°N)	Longitudinal (°E)
1	B	3.11	27.41	78.65
2	B	3.13	27.25	76.50
3	P	2.50	27.53	76.49
4	B	5.30	28.00	76.61
5	B	4.58	28.45	77.17

Table 8. Information of pothole (P) and bumps (B) collected by manual inspection for the R2 test site.

Scenario No.	Obstacle Type	Depth/Height (cm)	Latitude (°N)	Longitudinal (°E)
1	P	8.8	28.15	75.72
2	B	2.15	27.96	76.33
3	B	8.29	28.20	75.63
4	P	12.93	27.96	75.60
5	P	7.88	28.35	75.75
6	P	7.95	28.06	74.49
7	B	3.15	28.32	75.33
8	P	1.70	28.45	76.31
9	B	3.59	27.94	75.77
10	P	11.87	28.03	77.40

4.4. Validation of Proposed System with Other Techniques

The same experiment was conducted using three different methods several times on the very same roads; the rural road of Khora village and the urban road of the Yamuna Expressway. Each method, ultrasonic sensor, image processing, and manual inspection had slight deviations, as shown below. The comparison between three different techniques for depth/height, latitude, and longitude with their error are given in Tables 9–14 for both roads.

Table 9. Comparison of depth/height between the proposed system (PS), image processing (IP), and manual inspection (MI) for the R1 test site.

Scenario No.	Obstacle Type	Depth/Height (cm)				
		Proposed System	Image Processing	Manual Inspection	% Error (PS vs. IP)	% Error (PS vs. MI)
1	B	3.0	3.14	3.11	4.67	3.67
2	B	3.0	3.11	3.13	3.67	4.33
3	P	2.6	2.57	2.50	−1.15	−3.85
4	B	5.2	5.40	5.30	3.85	1.92
5	B	4.8	4.60	4.58	−4.17	−4.58

Table 10. Comparison of latitude between the proposed system (PS), image processing (IP), and manual inspection (MI) for the R1 test site.

Scenario No.	Obstacle Type	Latitude (°N)				
		Proposed System	Image Processing	Manual Inspection	% Error (PS vs. IP)	% Error (PS vs. MI)
1	B	28.4215	27.1937	27.41	4.32	3.56
2	B	28.4216	27.5405	27.25	3.1	4.12
3	P	28.4215	28.0691	27.53	1.24	3.14
4	B	28.4214	28.8080	28.00	−1.36	1.47
5	B	28.4214	27.9979	28.45	1.49	−0.1

Table 11. Comparison of longitude between the proposed system (PS), image processing (IP), and manual inspection (MI) for the R1 test site.

Scenario No.	Obstacle Type	Longitude (°E)				
		Proposed System	Image Processing	Manual Inspection	% Error (PS vs. IP)	% Error (PS vs. MI)
1	B	77.5261	76.5725	78.65	1.23	-1.45
2	B	77.5261	76.4872	76.50	1.34	1.32
3	P	77.5265	76.4799	76.49	1.35	1.34
4	B	77.5266	76.6505	76.61	1.13	1.18
5	B	77.5268	77.1004	77.17	0.55	0.46

Table 12. Comparison of depth/height between the proposed system (PS), image processing (IP), and manual inspection (MI) for the R2 test site.

Scenario No.	Obstacle Type	Depth/Height (cm)				
		Proposed System	Image Processing	Manual Inspection	% Error (PS vs. IP)	% Error (PS vs. MI)
1	B	9.27	8.8695	8.88	4.32	4.21
2	B	2.20	2.1318	2.15	3.10	2.34
3	P	8.63	8.3590	8.29	3.14	3.96
4	B	13.12	12.9271	12.93	1.47	1.47
5	B	8.20	8.2082	7.88	-0.10	3.96
6	P	8.19	7.8984	7.95	3.56	2.89
7	B	3.20	3.0682	3.15	4.12	1.63
8	P	1.77	1.7144	1.70	3.14	4.10
9	B	3.72	3.6010	3.59	3.20	3.50
10	P	12.16	11.6006	11.87	4.60	2.36

The deviation of the location of the pothole is relevant in different methods. The ultrasonic sensor and image processing detection deviated slightly from the actual position. Ergo, to calculate the accuracy of the system, the mean of the output using different method was calculated. The maximum error was 4.60% and the efficiency of the proposed system was about 95.48%.

Table 15 shows a full comparison between the results obtained with the proposed methodology and those yielded by similar approaches. As can be seen, our approach clearly outperformed the others, where TPR is the True Positive Rate, FPR is the False Positive Rate, and FNR is the False Negative Rate. One of the strongest points presented in this research is the use of a multi-variate model for pothole and speed bump detection. The presented methodology builds a model by combining different methods for the detection of road surface irregularities which provides greater accuracy to the entire framework. The collaboration of different methods with our novel framework targets a much bigger audience, and since it is very cost effective, it makes this system ideal for road surface monitoring in all regions.

Table 13. Comparison of latitude between the proposed system (PS), Image processing (IP), and Manual inspection (MI) for the R2 test site.

Scenario No.	Obstacle Type	Latitude (°N)				
		Proposed System	Image Processing	Manual Inspection	% Error (PS vs. IP)	% Error (PS vs. MI)
1	B	28.6186	27.4166	28.15	4.20	1.63
2	B	28.6243	27.4507	27.96	4.10	2.32
3	P	28.6543	27.6514	28.20	3.50	1.60
4	B	28.6722	28.3281	27.96	1.20	2.50
5	B	28.6932	27.7750	28.35	3.20	1.20
6	P	28.7423	28.2824	28.06	1.60	2.36
7	B	28.7865	28.0668	28.32	2.50	1.63
8	P	28.7933	28.0620	28.45	2.54	1.20
9	B	28.8654	27.9302	27.94	3.24	3.20
10	P	28.9653	28.2817	28.03	2.36	3.24

Table 14. Comparison of longitude between the proposed system (PS), image processing (IP), and manual inspection (MI) for the R2 test site.

Scenario No.	Obstacle Type	Longitude (°E)				
		Proposed System	Image Processing	Manual Inspection	% Error (PS vs. IP)	% Error (PS vs. MI)
1	B	77.5544	76.2903	75.72	1.63	2.36
2	B	77.5661	75.7666	76.33	2.32	1.60
3	P	77.5668	77.3884	75.63	0.23	2.50
4	B	77.5669	76.5430	75.60	1.32	2.54
5	B	77.5654	74.2999	75.75	4.21	2.34
6	P	77.5658	75.7508	74.49	2.34	3.96
7	B	77.5755	74.5035	75.33	3.96	2.89
8	P	77.5765	75.3345	76.31	2.89	1.63
9	B	77.5659	75.0760	75.77	3.21	2.32
10	P	77.5760	75.0858	77.40	3.21	0.23

Table 15. Detection rate comparison of different approaches on previous research.

Author	Detection Rate	Approach
Devapriya et al. [14]	30–92% TPR	Computer Vision Accelerometer & GPS
Eriksson et al. [8]	0.2 FPR	
Mohan et al. [7]	11.1% FPR & 22% FNR	Accelerometer, Microphone, GPS & GSM
Forrest et al. [10]	62% Accuracy	Ultrasonic
Singh et al. [11]	59.23% Accuracy	Global positioning system receiver and ultrasonic
Madli et al. [12]	74% Accuracy	Ultrasonic and GPS
Shivaleelavathi et al. [13]	79% Accuracy	Raspberry Pi, GPS, ultrasonic
Mohamed et al. [23]	75.6–87.8% (Accuracy)	Accelerometer
Astarita et al. [18]	90% Accuracy, 35% FP	Accelerometer using Smart Phone
Proposed system	95.48% (Accuracy)	Ultrasonic Sensor, Image Processing and manual Inspection

5. Conclusions

It has been highlighted that the current road anomaly detection frameworks used by the government or the existing authorities are very tedious and overwhelmingly time consuming, which is why this device eradicates all the problems and complications with the existing systems with much better efficiency. The analysis results demonstrated the use of ultrasonic sensors mated with an Arduino, GPS receivers, and HUD module, which runs on several filters. Dynamic time warping was also one of the algorithms used, which makes the validation process more manageable and is also one of the reasons for the increased efficiency. The presented Hanuman algorithm could detect the potholes and speed bumps accurately and efficiently. The calculated accuracy of the system was 95.48%, and the use of inexpensive components makes it the most cost effective framework among the existing systems. The model proposed in this paper serves two critical purposes (i.e., automatic detection of potholes, bumps, and alerting vehicle drivers to evade potential accidents). The HUD application provides real-time alerts about potholes and bumps right in front onto the windshield, making it more ergonomic. The framework works perfectly in all seasons, especially in monsoon as the signal/warning is generated from a database. The novel framework ha the highest efficiency when compared to all preexisting systems. Since it is a cost effective long time solution for pothole management, it can be easily provided to the maximum number of people; which can furthermore save many lives. Integration of GMaps and Satnav with the system can be done for ease of use. This framework can be applied to different types of road conditions and will be the future scope of this work.

Author Contributions: S.K.S. prepared the inspection system for the experiment and performed the experiments of this research. H.P. and J.L. conceived the original idea. J.L. designed the methodology and gave guidance and helped to improve the quality of the manuscript. S.K.S. wrote the original draft preparation in consultation with H.P. J.L reviewed and edited the final manuscript. All authors have read and agreed to the published version of the manuscript.

Funding: This work was supported by a National Research Foundation of Korea (NRF) grant funded by the Korea government (MSIT) (No. 2019R1A5A808320111, No. NRF-2019R1G1A1004577).

Acknowledgments: The authors would like to thank Amrit Seth, Ansh Mittal, and Vansh Malik, students at Amity University, who helped in conducting this research.

Conflicts of Interest: The authors declare that they have no conflict of interest. We declare that we do not have any commercial or associative interest that represents a conflict of interest in connection with the work submitted.

List of Abbreviations

GPS	Global Positioning System	PSD	Power Spectral Density
DTW	Dynamic Time Warping	UI	Unimportant Irregularity
HUD	Head Up Display	STD	Standard deviation
IFS	Irregularity Frame Selection	TPR	True Positive Rate
CIRC	Circularity	FPR	False Positive Rate
W	Average width	FNR	False Negative Rate

References

1. Malik, Y.S. *Road Accidents in India 2016*; Ministry of Road Transport & Highways: Delhi, India, 2016.
2. Guan, L.; Zou, M.; Wan, X.; Li, Y. Nonlinear Lamb Wave Micro-Crack Direction Identification in Plates with Mixed-Frequency Technique. *Appl. Sci.* **2020**, *10*, 2135. [CrossRef]
3. Schnebele, E.; Tanyu, B.F.; Cervone, G.; Waters, N. Review of remote sensing methodologies for pavement management and assessment. *Eur. Trans. Res. Rev.* **2015**, *7*, 7. [CrossRef]
4. Vittorio, A.; Rosolino, V.; Teresa, I.; Vittoria, C.M.; Vincenzo, P.G.; Francesco, D.M. Automated Sensing System for Monitoring of Road Surface Quality by Mobile Devices. *Procedia-Soc. Behav. Sci.* **2014**, *111*, 242–251. [CrossRef]
5. Praticò, F.G.; Fedele, R.; Naumov, V.; Sauer, T. Detection and Monitoring of Bottom-Up Cracks in Road Pavement Using a Machine-Learning Approach. *Algorithms* **2020**, *13*, 81. [CrossRef]
6. Vupparaboina, K.K.; Tamboli, R.R.; Shenu, P.M.; Jana, S. Laser-based detection and depth estimation of dry and water-filled potholes: A geometric approach. In Proceedings of the 2015 Twenty First National Conference on Communications (NCC), Mumbai, India, 27 February–1 March 2015; pp. 1–6.

7. Mohan, P.; Padmanabhan, V.N.; Ramjee, R. Nericell: Rich Monitoring of Road and Traffic Conditions using Mobile Smartphones. In Proceedings of the 6th ACM Conference on Embedded Network Sensor Systems, Raleigh, NC, USA, 5–7 November 2008; p. 323.
8. Eriksson, J.; Girod, L.; Hull, B.; Newton, R.; Madden, S.; Balakrishnan, H. The Pothole Patrol: Using a Mobile Sensor Network for Road Surface Monitoring. In Proceedings of the 6th International Conference on Mobile Systems, Applications, and Services, Breckenridge, CO, USA, 17-20 June 2008; p. 29. [CrossRef]
9. Kang, B.; Choi, S. Pothole detection system using 2D LiDAR and camera. In Proceedings of the 2017 Ninth International Conference on Ubiquitous and Future Networks (ICUFN), Milan, Italy, 4–7 July 2017; pp. 744–746.
10. Forrest, M.M.; Chen, Z.; Hassan, S.; Raymond, I.O.; Alinani, K. Cost Effective Surface Disruption Detection System for Paved and Unpaved Roads. *IEEE Access* **2018**, *6*, 48634–48644. [CrossRef]
11. Singh, G.; Kumar, R.; Kashtriya, P. Detection of Potholes and Speed Breaker on Road. In Proceedings of the 2018 First International Conference on Secure Cyber Computing and Communication (ICSCCC), Jalandhar, India, 15–17 December 2018; pp. 490–495.
12. Madli, R.; Hebbar, S.; Pattar, P.; Golla, V. Automatic Detection and Notification of Potholes and Humps on Roads to Aid Drivers. *IEEE Sens. J.* **2015**, *15*, 4313–4318. [CrossRef]
13. Shivaleelavathi, B.G.; Yatnalli, V.; Chinmayi, S.Y.V.; Thotad, S. Design and Development of an Intelligent System for Pothole and Hump Identification on Roads. *Int. J. Recent Technol. Eng.* **2019**, *8*, 5294–5300. [CrossRef]
14. Devapriya, W.; Babu, C.N.K.; Srihari, T. Real time speed bump detection using Gaussian filtering and connected component approach. In Proceedings of the 2016 World Conference on Futuristic Trends in Research and Innovation for Social Welfare (Startup Conclave), Coimbatore, India, 29 February 2016–1 March 2016; pp. 1–5.
15. Chen, K.; Lu, M.; Fan, X.; Wei, M.; Wu, J. Road condition monitoring using on-board Three-axis Accelerometer and GPS Sensor. In Proceedings of the 2011 6th International ICST Conference on Communications and Networking in China (CHINACOM), Harbin, China, 17–19 August 2011; pp. 1032–1037.
16. Arroyo, C.; Bergasa, L.M.; Romera, E. Adaptive fuzzy classifier to detect driving events from the inertial sensors of a smartphone. In Proceedings of the 2016 IEEE 19th International Conference on Intelligent Transportation Systems (ITSC), Janeiro, Brazil, 1–4 November 2016; pp. 1896–1901.
17. Aljaafreh, A.; Alawasa, K.; Alja'afreh, S.; Abadleh, A. Fuzzy inference system for speed bumps detection using smart phone accelerometer sensor. *J. Telecommun. Electron. Comput. Eng.* **2017**, *9*, 133–136.
18. Astarita, V.; Caruso, M.V.; Danieli, G.; Festa, D.C.; Giofrè, V.P.; Iuele, T.; Vaiana, R. A Mobile Application for Road Surface Quality Control: UNIquALroad. *Procedia-Soc. Behav. Sci.* **2012**, *54*, 1135–1144. [CrossRef]
19. Gonzalez, R.C.; Woods, R.E.; Masters, B.R. *Digital Image Processing*, 3rd ed.; Tata McGraw-Hill Education: New York, NY, USA, 2009; Volume 14, ISBN 0201180758.
20. Yu, B.X.; Yu, X. Vibration-Based System for Pavement Condition Evaluation. *Appl. Adv. Technol. Transp.* **2019**, *1*, 183–189.
21. Han, S.; Liu, M.; Shang, W.; Qi, X.; Zhang, Z.; Dong, S. Timely and Durable Polymer Modified Patching Materials for Pothole Repairs in Low Temperature and Wet Conditions. *Appl. Sci.* **2019**, *9*, 1949. [CrossRef]
22. Koch, C.; Brilakis, I. Pothole detection in asphalt pavement images. *Adv. Eng. Inform.* **2011**, *25*, 507–515. [CrossRef]
23. Mohamed, A.; Fouad, M.M.M.; Elhariri, E.; El-Bendary, N.; Zawbaa, H.M.; Tahoun, M.; Hassanien, A.E. *RoadMonitor: An Intelligent Road Surface Condition Monitoring System*; Springer: Cham, Switzerland, 2015; pp. 377–387.

Article

Defects Inspection in Wires by Nonlinear Ultrasonic-Guided Wave Generated by Electromagnetic Sensors

Junpil Park [1], Jaesun Lee [2], Junki Min [3] and Younho Cho [1,*]

1 School of Mechanical Engineering, Pusan National University, Busan 46241, Korea; jpp@pusan.ac.kr
2 School of Mechanical Engineering, Changwon National University, Changwon 51140, Korea; jaesun@changwon.ac.kr
3 Gimhae Industry Promotion & Bio-Medical Foundation, Gimhae 50989, Korea; mjk6412@pusan.ac.kr
* Correspondence: mechcyh@pusan.ac.kr; Tel.: +82-51-510-2323

Received: 26 May 2020; Accepted: 25 June 2020; Published: 28 June 2020

Featured Application: The second harmonic measurement method using EMAT proposed in this paper has established a system that can detect microdefects in wires without destroying objects before wire processing.

Abstract: Steel wires are widely used as raw materials for spring valves in engines. Considering the quality and safety issues of their structure, there is a demand to develop nondestructive inspection approaches to detect initial damages in steel. In this study, nonlinear ultrasonic-guided waves generated by an electromagnetic acoustic transducer (EMAT) were used to inspect the defects in steel wires. As one of the noncontact testing methods, the use of EMAT has significant advantages to decrease the nonlinearity induced by instruments and transducer contact condition. The principles of design and manufacturing of EMAT are first introduced. The fundamental theory of nonlinear guided waves is also briefly discussed in this investigation. Phase-matched guided wave modes were generated and measured by using EMAT. Variations of acoustic nonlinearity corresponding to existing defects in specimens were obtained. A scanning electron microscope (SEM) was used to check the existence of microdefects in specimen. The results indicate that the use of EMAT can be an effective means to generate and measure nonlinear ultrasonic-guided waves for inspection of microdefects.

Keywords: steel wire rod; nonlinear ultrasonic; EMAT; guided wave

1. Introduction

Steel wire rods are used as raw materials for various mechanical components. This research can be applied to the steel wires, raw material of valve springs and have an influence on keeping smooth cycle motion in a gasoline engine. Under periodic fatigue state, failure of spring valves may occur before expectation of fatigue failure. Therefore, we need a reliable method for steel wires that can evaluate microdefects or early stage damage detection. Up to now, for the research work related with this study not so many articles were published. Gao et al. developed nondestructive evaluation (NDE) techniques using magnetic flux for macro defect [1].

The geometric shape of steel wires before processing of valve springs is cylindrical such as rod and solid shaft. Considering shape characteristics, guided waves have been applied to efficiently inspect long specimens such as wire rods. Research on guided ultrasound has been going on for a long time, but it has not progressed actively due to the complexity of theory and the variety of wave modes. Gazis performed basic guided ultrasonic analysis of cylindrical structures [2,3]. Based on this, the test for guided ultrasound was started by Rose and Cawley's research works [4,5].

The nonlinear technique has the advantage that it can examine the microstructural changes in the nonlinear elastic section. It is based on information of distance, time and velocity of ultrasonic wave used in linear technique. In addition, the acquired ultrasonic signal is further analyzed in the frequency domain. The acquisition method of this technique is as follows: Information on the defect can be obtained by comparing the change at the second or third harmonic frequency. Through this, the internal microdefect can be diagnosed early. It is also possible to predict the life expectancy based on quantitative results. Buck et al. [6] conducted a study on the change of function of the stress and the second harmonic in which the transition occurs in the material. Subsequently, Jacobs and colleagues [7–9] studied the plasticization of materials using nonlinear guided ultrasound. Furthermore, according to using acoustic nonlinear guided wave technique, phase-matching mode is required to generate higher order harmonic amplitude. The basic concept of this research, concerning which factor is related with phase velocity and group velocity, has been improved by M. Deng [10,11].

EMAT is a noncontact technique. Research on EMAT has been conducted by M. Hiaro and H. Ogi of Osaka University [12–14]. Quantitative changes of physical nonlinearity should be measured by ultrasonic nonlinear method. Therefore, there is a need to continuously carry out experiments under the same conditions. That is, the same experimental apparatus and conditions are required to minimize the influence of the nonlinearity of the system. At this time depending on the contact conditions of the excitation and receiving transducers the reliability of this experiment may be greatly affected. This may change the nonlinearity. Therefore, to overcome the nonlinearity, a noncontact EMAT was applied.

Currently, research on microdefects in the material is proceeding smoothly by combining the contact method and the nonlinear method. However, the contact test method using piezoelectric transducer (PZT) requires a certain size of contact area, but it is impossible to test using PZT in a circular form in the case of the wire rod studied in this study. In addition since the PZT patch requires a contact medium such as an ultrasonic couplant, the signal is irregular depending on the amount or contact strength of the contact medium, so reproducibility is not constant when performing repeated experiments. Therefore, research using noncontact technology is necessary, but the result of the research is insufficient. In particular, there have been no reports on the application of the noncontact technique to the internal microdefects of steel wire rods. Therefore, this study applies the nonlinear method, guided wave and noncontact method.

Steel wire rod, which is the raw material of the engine piston valve spring, is typically manufactured from three materials. The nonlinear tendency according to the property of each material was analyzed to secure the suspected defective area. Flexible, torsional and longitudinal modes exist for guided waves in cylindrical structures such as piping and wire rods. Longitudinal mode is applied to ensure smooth secondary harmonics. In addition, we tried to obtain the reliability of the experimental results by setting the same geometric conditions. The nonlinear tendency of each specimen can be grasped through experiments. According to the result of the experiment, the suspected defect area was secured. This study provides a technical and academic understanding of nonlinear guided wave. By using noncontact ultrasonic technique rather than contact acoustic nonlinear guided waves, we can study for the purpose of precise diagnosis.

2. Nonlinear Guided Waves

2.1. Nonlinear Ultrasonic Test

The method for detecting nonlinearity in a material is to send and receive induced ultrasonic wave at a specific distance from the material. After the nonlinearity of the material propagates at a certain distance, the underlying frequency wave is distorted to produce a higher harmonic [15–21]. For guided wave, the acoustic nonlinearity can be written as shown in Equation (1).

$$\beta = \frac{8}{K^2 a}\frac{A_2}{{A_1}^2}f \tag{1}$$

where a is the wave propagation distance and f is a frequency independent function. In addition, the ratio A_2/A_1^2 with the relative acoustic nonlinear parameter β' applied in this research is considered as a measure of acoustic nonlinearity, where A_1 and A_2 are the measured amplitudes of the fundamental and second harmonics.

$$\beta' = \frac{A_2}{A_1} \propto \beta a \quad (2)$$

The relative acoustic nonlinear parameter increases linearly as a function of propagation distance. The cumulative effect of the second harmonic amplitude is a great advantage for detection in experiment. It is found that second harmonic amplitude linearly grows with propagation distance, when there are synchronism and nonzero power flux. If the wave mode chosen satisfies these two conditions, the second harmonic amplitude will be cumulative. Even a series of double frequency wave components will be generated by the driving sources of nonlinearity. In practice, interest is focused on the second harmonic generation with the cumulative effect, since the cumulative second harmonic wave plays a dominant role after second harmonics propagating some distance.

Figure 1a,b shows the typical waveform of the received signal and its spectrum. The received time domain signal is processed in the frequency domain with the fast Fourier transform (FFT) to obtain its spectrum. Figure 1b shows the fundamental frequency and the second harmonic.

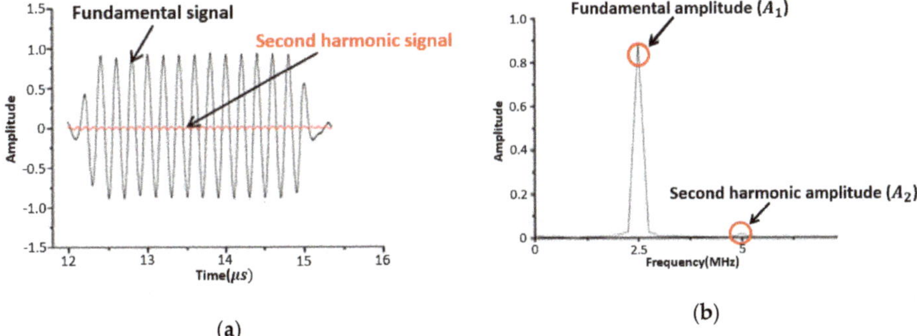

Figure 1. Typical received signal. (**a**) Time-domain waveform; (**b**) Fourier spectra of the fundamental and second harmonic signal.

As indicated in Equations (1) and (2), the normalized nonlinear parameter can be calculated with Equation (2) by using the amplitudes of the fundamental and the second harmonic modes.

Figure 2 shows the nonlinear parameters according to the propagation distance. The cumulative effect means that the nonlinearity increases with distance under certain frequency conditions. The double frequency mode which satisfies a specific condition such as the expression shown in the diagram is selected based on the frequency.

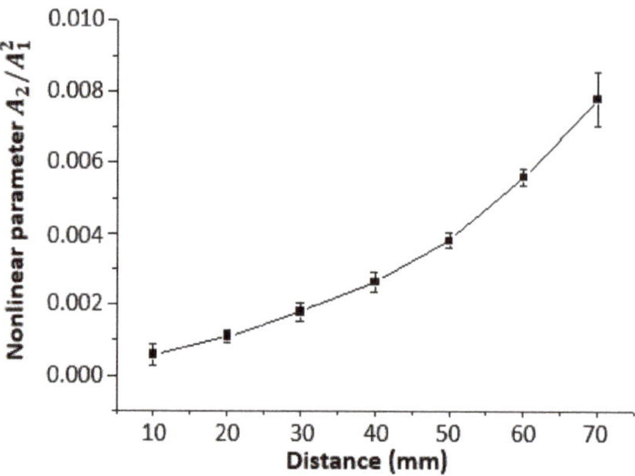

Figure 2. Nonlinear parameters according to distance.

2.2. EMAT

EMAT is composed of a generating coil for generating an eddy current of the electromagnet and the transmission and a receiving detecting coil. When the probe is approached to the metal specimen, the inside of the specimen is affected by the electromagnet. As Shown in Figure 3, the eddy current is formed on the surface of the specimen by an alternating current flowing in the generating coil inside the probe. Lorentz force is generated between the eddy current and the low power line, and mechanical displacement occurs on the surface of the specimen. This mechanical displacement is called ultrasound.

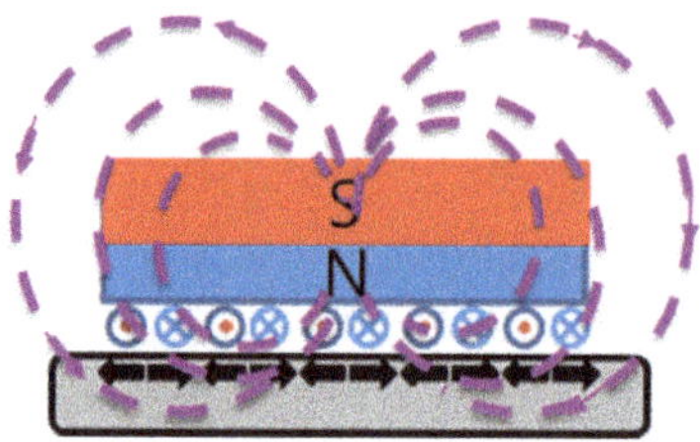

Figure 3. Principle of electromagnetic acoustic transducer (EMAT) with magnetic flux.

The purpose of this study was to investigate the nonlinearity of materials by inducing an ultrasonic wave in the longitudinal mode in steel wire rod. The permanent magnet was magnetically biased in the axial direction on the specimen. Then, when the coil was wound in the circumferential direction and the current was applied, the longitudinal mode of the guided ultrasonic wave was generated in the cylindrical structure such as the pipe or the wire rod.

EMAT was generated by Lorentz force effect in meander coil. The alternative current induce in the coil and the magnetic field is built in coil by high-power permanent magnet. Here **J** is eddy currents field and B is the magnetic field. The Lorentz force is generated by the eddy currents and magnetic field

$$\vec{F} = \vec{J} \times \vec{B} \tag{3}$$

As shown in Figure 4, the geometry used for describing the reception of EMAT is considering various factors. Equation (3) explain the relationship between magnetic field and eddy current. The influence of magnetic field for inducing eddy current is indicated by Equations (4) and (5). Produce a voltage pick up by receiving coil is Equation (6). EMAT does not require a contact medium in flaw detection. As a noncontact method using electromagnetic, it shows higher performance than

PZT for measuring nonlinearity of materials. However, it should be considered that the receiving sensitivity is lower than PZT.

$$\begin{array}{l} E_{ed} = v \times B \\ J_{ed} = \sigma \times E_{ed} \end{array} \quad \vec{F} = \vec{J} \times \vec{B} \tag{4}$$

$$\begin{array}{l} \nabla^2 A_0 = j\omega\sigma_0\mu_0 A_0 \\ \nabla^2 A_1 = -\omega^2\sigma_1\mu_1 A_1 \\ \nabla^2 A_2 = j\omega\sigma_2\mu_2 A_2 \\ \nabla^2 A_3 = j\omega\sigma_3\mu_3 A_3 \end{array} \quad \vec{F} = \vec{J} \times \vec{B} \tag{5}$$

$$\begin{array}{l} E_r = -\frac{\partial A}{\partial t} \\ V_r = N \times l \times E_r \end{array} \quad \vec{F} = \vec{J} \times \vec{B} \tag{6}$$

where v is the velocity of particle vibration, B is the static magnetic field, σ is the conductivity of the test piece, E_{ed} and $\mathbf{J}_{ed}$ are the induced electric field and the induced eddy current sheet due to the particle vibration at the depth z_{ed} respectively.

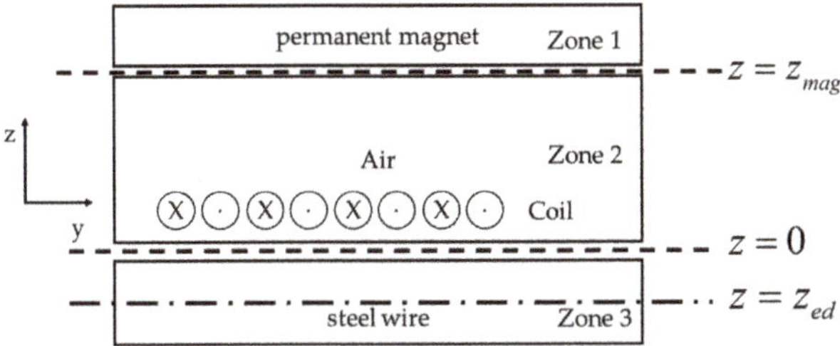

Figure 4. Geometry of steel wire and EMAT.

3. Experiments

Three different kinds of specimens with total 9 samples were tested in this investigation. As shown in Figure 5, the size information of the three specimens was the same, but there were differences in material properties and tensile strength. Detailed information of the samples can be found in in Table 1.

Figure 5. Specimens of steel wire rods.

Table 1. Material properties of specimens.

List	Material Name		
	SWOSC-V	SWOSC-VHV	SWOSC-VHS
Line diameter		3.2 mm	
Length		322 mm	
Density		787.4 g/mm^3	
Modulus of elasticity		206 kN/mm^2	
Yield stress		0.9 × Tensile Strength	
Tensile strength	1860–2010 N/mm^2	2010–2160 N/mm^2	2110–2210 N/mm^2
C content	0.51–0.59	0.63–0.68	0.63–0.68
Si content	1.20–1.60	1.20–1.60	1.80–2.20
Mn content	0.50–0.80	0.50–0.80	0.70–0.90

The frequency and modes were selected by frequency multiply wire thickness ($f \times d$). Phase matching must be performed through the dispersion curve. The conditions for generation of second harmonic in a cylindrical structure such as wire rod and pipe are as follows [13].

The power flux at the first harmonic should not be 0 in the second harmonic frequency range. $({f_n}^{surf} + {f_n}^{vol} \neq 0)$

Phase and group velocities of the first and second harmonics should be the same through phase matching (${k_n}^* = 2k$).

Figure 6 shows the dispersion diagram of steel wire rod. It can be confirmed that the phase velocity and the group velocity were the same at 4 MHz · mm, which was a double frequency of 2 MHz · mm. Through this phase-matching, we could use a nonlinear guided wave technique. L_1 (f·d = 2 MHz·mm) and L_2 (f·d = 4 MHz·mm) were selected because the resulted of the L_1 and L_2 modes were clearer for the second harmonic component than the other modes.

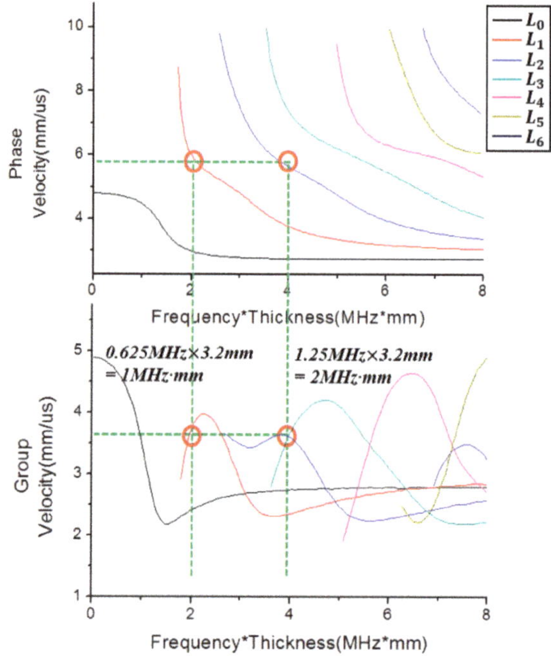

Figure 6. Dispersion curve of steel wire rod.

EMAT was designed and manufactured through impedance and LC-matching. Impedance matching means to eliminate the phase difference by minimizing reflection or loss of input power by buffering impedance difference at input and output. Selecting the appropriate inductor and capacitor element values through Equation (7) and choosing the resonance frequency is called LC-matching.

$$Z = R + jwL + \frac{1}{jwC}\vec{F} = \vec{J} \times \vec{B} \quad (7)$$

The frequency at which the imaginary part becomes 0 in the Equation (7) through matching between the inductor (L) and the capacitor (C) is called a resonance frequency. Figure 7 shows the resonance frequency due to LC-matching of the capacitor and the inductor.

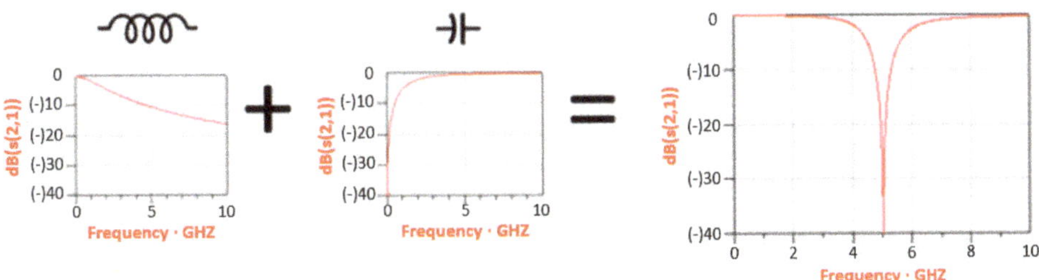

Figure 7. Resonant frequency setting by LC-matching of capacitor and inductor.

In order to realize the inductance in the ultrasonic generating coil, a long meander coil was used in a limited space. A resonant circuit was constructed to maximize current flow in the meander coil. Figure 8 shows a meander coil with a wavelength of 0.38 mm.

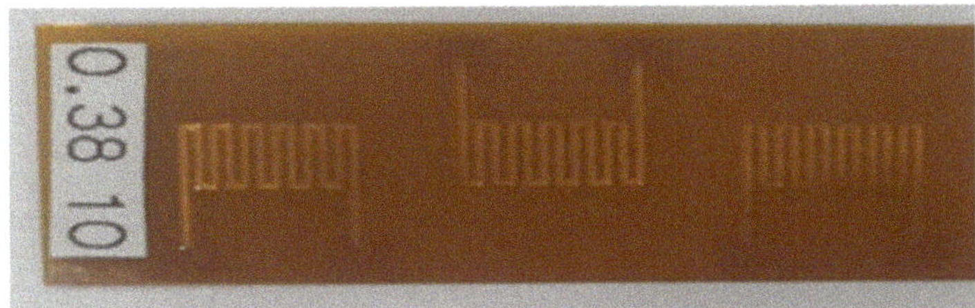

Figure 8. Meander coils with a wavelength of 0.38 mm.

Figure 9 shows the impedance-matching EMAT transducer used in this study. The EMAT was manufactured at 1 MHz as shown in Figure 10 and excited at 0.625 MHz. The wavelength (λ = 5.82 mm) of 1 MHz can be derived by substituting the desired frequency and longitudinal wave velocity ($C^L{}_{Steel} = 5.82\,\mathrm{mm}/\mu\mathrm{s}$) in the steel into Equation (8).

$$\lambda = \frac{c}{f}\vec{F} = \vec{J} \times \vec{B} \tag{8}$$

Figure 9. Impedance-matched EMAT transducer.

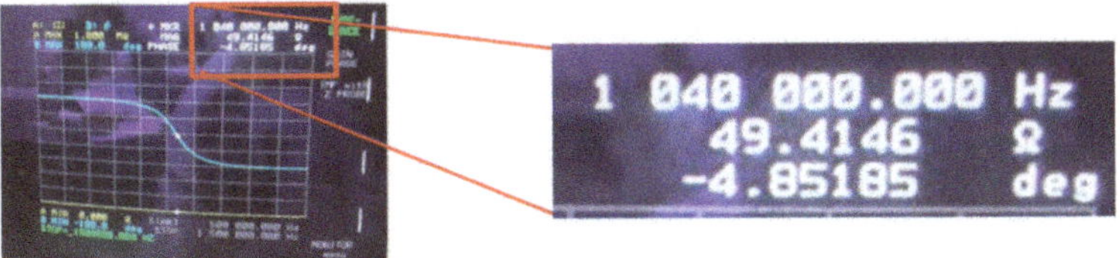

Figure 10. Impedance matching.

Practically, 5.82 mm cannot be applied to steel wire with a diameter of 3.2 mm. It can be applied with lower wavelength as $\lambda/2, \lambda/4, \lambda/8$. Since 0.38 mm is an approximate value of 1/16 of wavelength, EMAT was fabricated using 0.38-mm meander coil.

Experiments were conducted to investigate the behavior of nonlinear induced ultrasound. Figure 11 shows the experimental setup. The RITEC RAM-4000 was used as a tone-burst ultrasonic transmitter and receiver and it was processed by a pitch–catch method.

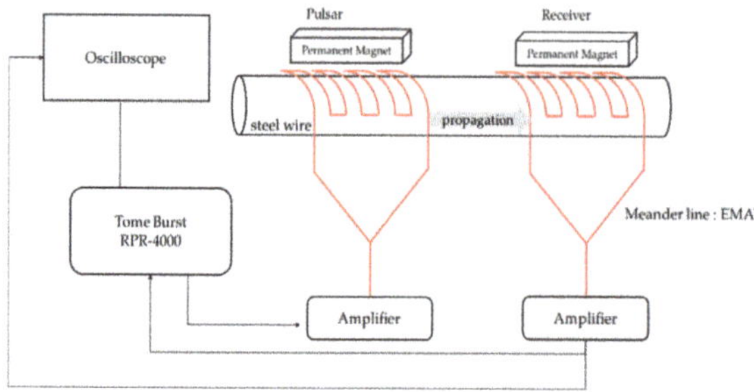

Figure 11. Experimental setup for using EMAT.

In order to obtain the reliability of the experimental results, the same geometric conditions were set, and two kinds of experiments were conducted. As shown in Figure 12, a meander line and permanent magnets were installed on the steel wire rod. As Shown in Figure 13a, the first experiment was divided into four sections of 60 mm each, and the relative nonlinearity of each wire was confirmed. The second experiment was conducted by increasing the propagation distance by 60 mm as shown in Figure 13b. The schematic diagram of each experiment is shown in Figure 13. Experiments were conducted to determine whether defects could be detected through the relative nonlinearity.

Figure 12. Setup of the meander line and permanent magnet on steel wire rod.

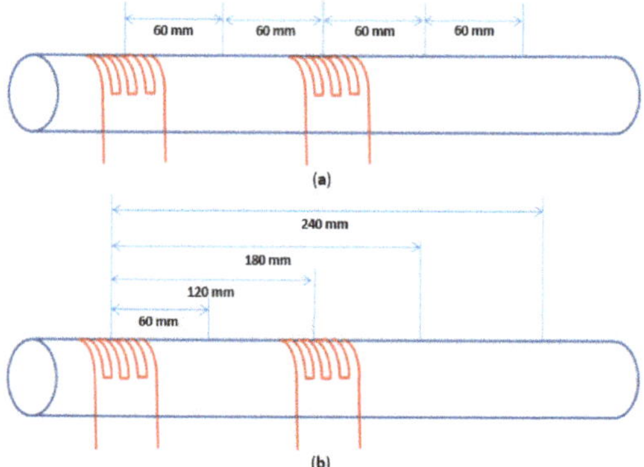

Figure 13. Schematic of experimental: (**a**) Experimental method with each detection area; (**b**) experimental method with increasing distance.

4. Results and Discussion

The relative nonlinearity of the fabricated EMAT sensor was measured for each specimen in the longitudinal direction of the steel wire. The relative nonlinearity tendency was studied. As mentioned above, the nonlinear guided wave technique is a technique to diagnose changes in material properties and microdefects by using harmonic components unlike the conventional technique. EMAT shows higher performance than PZT to measure nonlinearity of material because it does not need contact medium and uses electromagnetic in flaw detection. In case of using PZT, the error due to the individual difference of the experimenter cannot be ignored. Therefore, the method using EMAT is very appropriate. Figure 14 shows the signals for linear ultrasonic wave in three specimens. The time domain signal for each specimen can distinguish between the distance and the speed relative to the time, however, microdefects cannot be confirmed. In addition, the analysis of the modes showed that the L_1 mode was well implemented for SWOSC-V and SWOCS-VHV, but SWOSC-VHS was able to confirm that the various modes were mixed. Therefore, the experiment was conducted using a nonlinear guided ultrasonic technique using second harmonic.

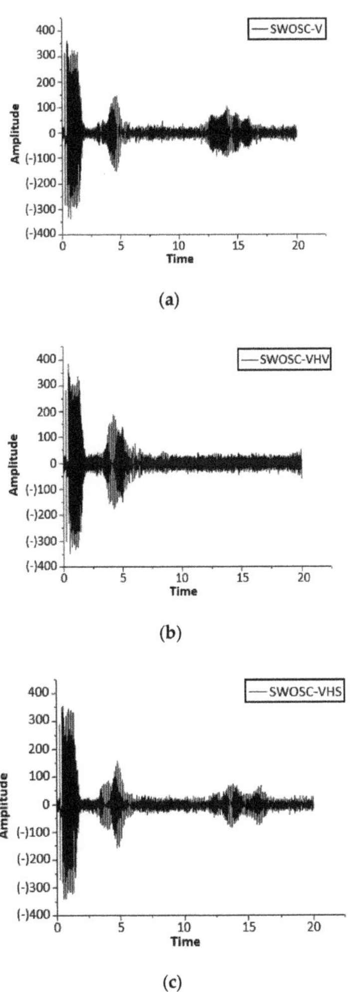

Figure 14. Time domain signal. (**a**) SWOSC-VHS #1; (**b**) SWOSC-VHV #2; (**c**) SWOSC-VHS #3.

The fundamental amplitude A1 and second harmonic amplitude A2 amplitude values can be derived by measuring the time domain signal of 60 mm intervals from three specimen for each material and then performing FFT. Substituting the derived A1 and A2 into Equation (2), nonlinear parameters can be obtained for each specimen. Figure 15 is the result of comparing the relative nonlinearity in each zone. In all three specimens of SWOSC-V, the relative nonlinearity increases with increasing propagation distance. The relative nonlinearity of all three specimens decreases at the propagation distance of 181–240 mm, which can be regarded as the material properties of the specimen. SWOSC-VHV also shows the same tendency of three specimens. However, SWOSC-VHS # 2 has a tendency to be different from the # 1 and # 3 specimens at 61–180 mm propagation distance. Three specimens with the same physical properties should show the same tendency toward in each section. However, since VWOSC-VHS #2 has a different tendency from the VWOSC-VHS #1 and #3 samples, it can be determined that there is some abnormality inside the material.

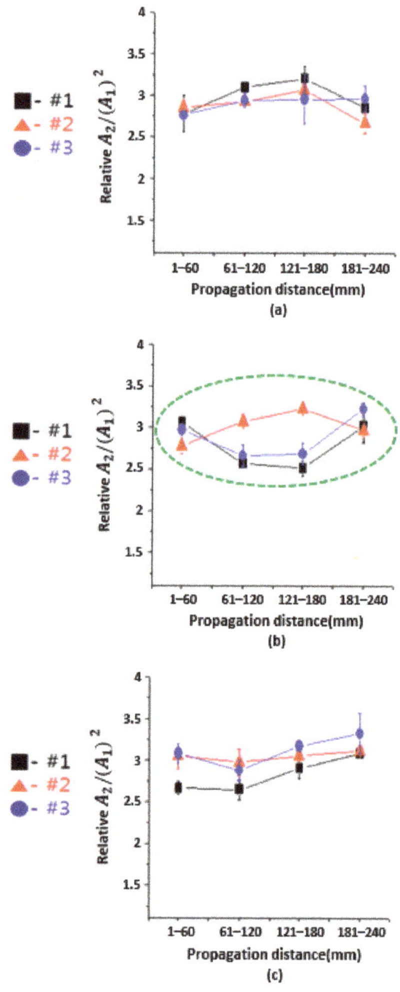

Figure 15. Relative nonlinearity for each detection area (**a**) Results of the three-specimen test in SWOSC-V (**b**) results of the three-specimen test in SWOSC-VHV (**c**) results of the three-specimen test in SWOSC-VHS.

Because of the tendency to suspect microscopic defects, additional experiments were conducted to ensure the reliability of the experimental data. Figure 16 is the relative nonlinearity with increasing distance, which was an additional experimental result. SWOSC-V and SWOSC-VHS could obtain a tendency for relative nonlinearity change as the distance increased. However, # 3 of SWOSC-VHV showed a large difference in relative nonlinear factor values from # 1 and # 2 at a propagation distance of 120 mm, but it was very difficult to be considered as a suspected defect region due to the agreement of the tendency for each specimen. At the propagation distance of 180 mm, SWOSC-VHH # 2 had the same tendency as the previous experimental results. A total of two experiments were used to consider suspected microdefects in SWOSC-VHV # 2 at a propagation distance of 180 mm.

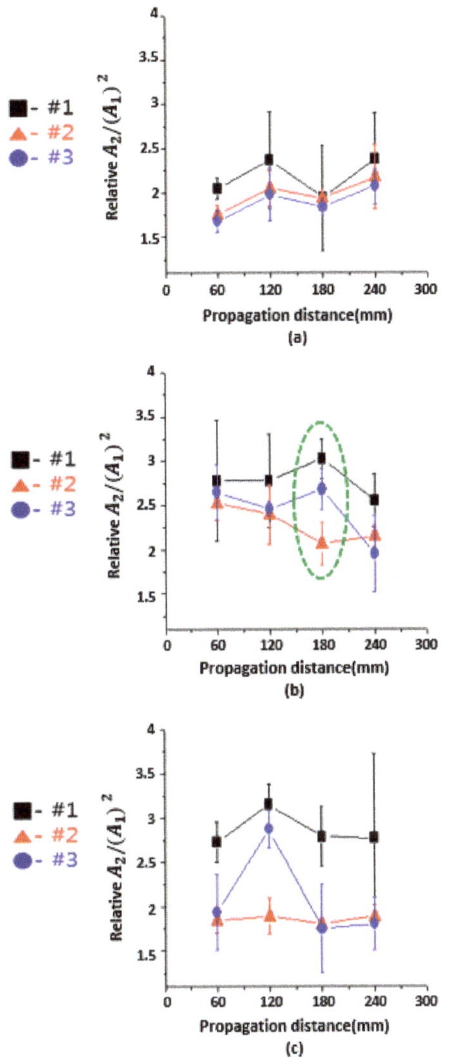

Figure 16. Relative nonlinearity variation of considering propagation distance (**a**) Results of the three-specimen test in SWOSC-V (**b**) results of the three-specimen test in SWOSC-VHV (c) results of the three-specimen test in SWOSC-VHS.

Figure 17 is cross-sectional photographs of microdefect suspected areas and defective free area. Figure 17b shows a clear difference in the distribution of microdefects compared to Figure 17a. This is consistent with the experimental results and SEM photographs. Therefore, it can be seen that the reliability of the experimental result was secured. Figure 18 shows the result of enlarged the micro defect.

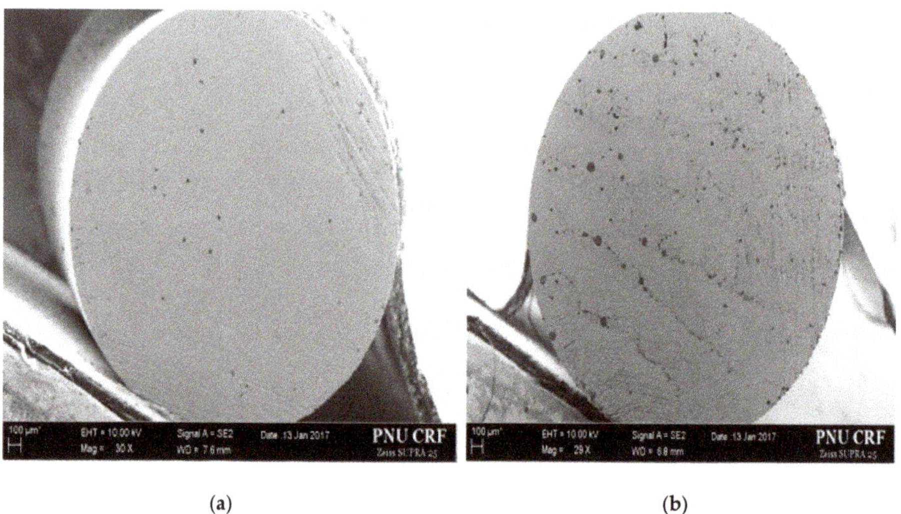

(**a**) (**b**)

Figure 17. Comparison of SEM result for steel wire. (**a**) SWOSC-VHS #1: non-defect area; (**b**) SWOSC-VHS #2: defect area.

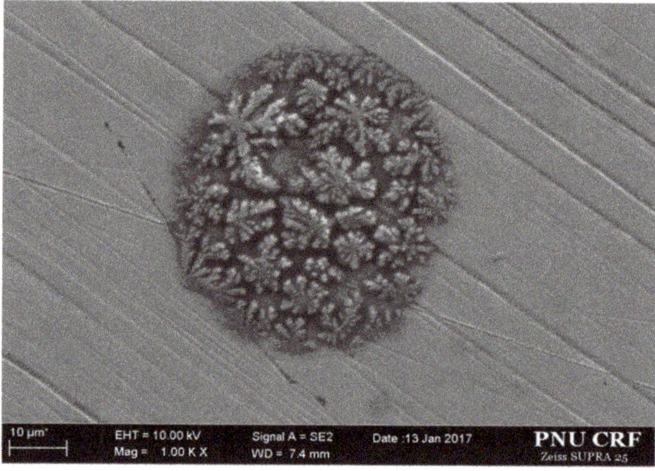

Figure 18. Enlarged microdefect.

The experimental results were not quantitative because the distance between the coils and the specimen changed finely when the reinstallation was repeated after separating the meander coil and the specimen from each other during the experiment. In other words, the lift-off changed. In order to solve this problem, we had to build a proper system to fix the meander coil and the specimen firmly. In addition, since the nonlinearity was very sensitive, the ultrasound path difference may have occurred due to the minute scratches on the surface of the specimen, resulting in energy loss and affecting the damping coefficient. Therefore, it is necessary to acquire precision machined specimens.

5. Conclusions

In this study, the nonlinearity tendency of microdefects was investigated by applying guided wave and nonlinear method to steel wire using EMAT. The relative nonlinearity of each specimen is specific for each material. This is considered to be a characteristic of the content of elements constituting the material. Experimental results show that specimens with large errors and specimens without specimens exist. This is considered to be an error due to the distance between the meander coil and the specimen, that is, the lift-off, when the specimen and the coil are separated and remounted. SEM images were taken to verify the reliability of the experimental results. The results were very satisfactory. The possibility of diagnosing microdefects in steel wire production line was confirmed. In order to utilize the nonlinear technique, there is a need to acquire defect-free specimens. We have identified the need for a system that can acquire clear signals. However, there were experimental errors, there is no critical dispute in confirming the tendency for each specimen. It is possible to observe relatively the suspicion of microdefect. It is expected that quantitative evaluation of each specimen will be possible if the parameters of nonlinear factor are removed through a system which can fix the defect-free specimen and lift-off later. In addition, if this research technique is used, it is possible to quickly check noncontact for defects of steel wire rods passing through the inspection line, and to detect fine defects early before processing the steel wire rod to enhance the safety of the material. It was suitable for inspecting the initial fine defects in steel wires before spring processing by utilizing EMAT, a noncontact method without contact medium. The noncontact method is a method capable of minimizing the deformation of the signal that changes depending on the amount or compression state of the contact medium. If the technique is applied to the industrial site, it is judged that it will be possible to test for microdefects quickly and consistently.

Author Contributions: J.P. and J.M. designed and performed the experiments. J.L. and Y.C. conceived the original idea. J.P. developed the theory and performed the computations with support from J.L. and Y.C. verified the analytical methods. J.M. and J.L. wrote the draft study, J.P. completed the final study and Y.C. made the final review. All authors have read and agreed to the published version of the manuscript.

Funding: This work was supported by the Korean Evaluation Institute of Industrial Technology (KEIT) grant funded by the Korean government (MOTIE) (NO.10085576).

Acknowledgments: This research was carried out with the support of materials from Daewon Kang Up Co., Ltd.

Conflicts of Interest: The authors declare no conflict of interest.

References

1. Yan, X.; Zhang, D.; Pan, S.; Zhang, E.; Gao, W. Online nondestructive testing for fine steel wire rope in electromagnetic interference environment. *NDT E Int.* **2017**, *90*, 75–81. [CrossRef]
2. Gazis, D.C. Three dimensional investigation of the propagation of waves in hollow circular cylinders. *J. Acoust. Soc. Am.* **1959**, *31*, 568–573. [CrossRef]
3. Rose, J.L. *Ultrasonic Waves in Solid Media;* Cambridge University Press: Oxford, UK, 2004.
4. Rose, J.L. A baseline and vision of ultrasonic guided wave inspection potential. *J. Pres. Ves. Tech.* **2002**, *124*, 273–282. [CrossRef]
5. Lowe, M.J.S.; Alleyne, D.N.; Cawley, P. Defect detection in pipes using guided waves. *Ultrasonics* **1998**, *36*, 147–154. [CrossRef]
6. Buck, O. Harmonic generation for measurement of internal stress as produced by dislocation. *IEEE Trans. Sonics Ultrason.* **1976**, *SU-23*, 346–350. [CrossRef]
7. Shui, G.; Kim, J.Y.; Qu, J.; Jacobs, L. A new technique for measuring the acoustic nonlinearity of material using Rayleigh waves. *NDT E Int.* **2008**, *41*, 326–329. [CrossRef]
8. Kim, J.Y.; Qu, J.; Jacobs, L.J. Acoustic nonlinearity parameter due to micro-plasticity. *J. Nondestruct. Eval.* **2006**, *25*, 29–37. [CrossRef]

9. Pruell, C.; Kim, J.-Y.; Qu, J.; Jacobs, L.J. Evaluation of plasticity driven material damage using Lamb waves. *Appl. Phys. Lett.* **2007**, *91*, 231911. [CrossRef]
10. Deng, M. Cumulative second harmonic generation of Lamb mode propagation in a solid plate. *J. Appl. Phys.* **1999**, *85*, 3051–3058. [CrossRef]
11. Deng, M. Analysis of second-harmonic generation of Lamb waves propagating in layered planar structures with imperfect interfaces. *Appl. Phys. Lett.* **2006**, *88*, 221902. [CrossRef]
12. Hirao, M.; Ogi, H. Electromagnetic acoustic resonance and materials characterization. *Ultrasonics* **1977**, *35*, 413–421. [CrossRef]
13. Hirao, M.; Ogi, H.; Fukuoka, H. Resonance EMAT system for acoustoelastic stress measurement in sheet metals. *Rev. Sci. Instruments* **1993**, *64*, 3198–3205. [CrossRef]
14. Hirao, M.; Ogi, H. An SH-wave EMAT technique for gas pipeline inspection. *NDT E Int.* **1999**, *32*, 127–132. [CrossRef]
15. Li, W.; Deng, M.; Xiang, Y.X. Review on second-harmonic generation of ultrasonic guided waves in solid media (I): Theoretical analyses. *Chin. Phys. B* **2017**, *26*, 114302. [CrossRef]
16. Matlack, K.H.; Kim, J.-Y.; Jacobs, L.J.; Qu, J. Experimental characterization of efficient second harmonic generation of lamb wave modes in a nonlinear elastic isotropic plate. *J. Appl. Phys.* **2011**, *109*, 014905. [CrossRef]
17. Li, W.; Deng, D.; Cho, Y. Cumulative Second Harmonic Generation of Ultrasonic Guided Waves Propagation in Tube-Like Structure. *J. Comput. Acoust.* **2016**, *24*, 1650011. [CrossRef]
18. Kim, J.; Zhu, B.; Cho, Y. An experimental study on second harmonic generation of guided wave in fatigued spring rod. *J. Mech. Sci. Technol.* **2019**, *33*, 4105–4109. [CrossRef]
19. Li, W.; Jiang, C.; Qing, X.; Liu, L.; Deng, M. Assessment of low-velocity impact damage in composites by the measure of second-harmonic guided waves with the phase-reversal approach. *Sci. Prog.* **2019**, *103*, 103. [CrossRef] [PubMed]
20. Li, W.; Chen, B.; Qing, X.; Cho, Y. Characterization of Microstructural Evolution by Ultrasonic Nonlinear Parameters Adjusted by Attenuation Factor. *Metals* **2019**, *9*, 271. [CrossRef]
21. Li, W.; Cho, Y.; Li, X. Comparative Study of Linear and Nonlinear Ultrasonic Techniques for Evaluation Thermal Damage of Tube-Like Structures. *J. Korean Soc. Nondestruct. Test.* **2013**, *33*, 1–6. [CrossRef]

Article

High-Precision Noncontact Guided Wave Tomographic Imaging of Plate Structures Using a DHB Algorithm

Junpil Park [1], Jaesun Lee [2], Zong Le [3] and Younho Cho [1,*]

1 School of Mechanical Engineering, Pusan National University, Busan 46241, Korea; jpp@pusan.ac.kr
2 School of Mechanical Engineering, Changwon National University, Changwon 51140, Korea; jaesun@changwon.ac.kr
3 National Deep Sea Center, Ministry of Natural Resources, Qingdao 266247, China; zongl@ndsc.org.cn
* Correspondence: mechcyh@pusan.ac.kr; Tel.: +82-51-510-2323

Received: 26 May 2020; Accepted: 24 June 2020; Published: 25 June 2020

Abstract: The safety diagnostic inspection of large plate structures, such as nuclear power plant containment liner plates and aircraft wings, is an important issue directly related to the safety of life. This research intends to present a more quantitative defect imaging in the structural health monitoring (SHM) technique by using a wide range of diagnostic techniques using guided ultrasound. A noncontact detection system was applied to compensate for such difficulties because direct access inspection is not possible for high-temperature and massive areas such as nuclear power plants and aircraft. Noncontact systems use unstable pulse laser and air-coupled transducers. Automatic detection systems were built to increase inspection speed and precision and the signal was measured. In addition, a new Difference Hilbert Back Projection (DHB) algorithm that can replace the reconstruction algorithm for the probabilistic inspection of damage (RAPID) algorithm used for imaging defects has been successfully applied to quantitative imaging of plate structure defects. Using an automatic detection system, the precision and detection efficiency of data collection has been greatly improved, and the same results can be obtained by reducing errors in experimental conditions that can occur in repeated experiments. Defects were made in two specimens, and comparative analysis was performed to see if each algorithm can quantitatively represent defects in multiple defects. The new DHB algorithm presented the possibility of observing and predicting the growth direction of defects through the continuous monitoring system.

Keywords: tomographic imaging; DHB algorithm; noncontact; air-coupled transducer

---

1. Introduction

It is well known that Lamb waves can propagate long distances with little energy loss. Lamb waves have been shown to be sensitive to most types of defects. They are also efficient for detecting tiny subsurface damages and inspecting large areas. Such noncontact nondestructive testing (NDT) methods have great advantages for online inspection because high temperature or other negative factors can restrict the application of contact methods. In earlier research, ultrasonic Rayleigh and Lamb waves were used to reconstruct images in metal plates using water immersion techniques [1]. Other studies have used electromagnetic acoustic transducer (EMAT)-based methods [2–4] and contact techniques [5–9]. Air-coupled transducers have been used for tomography imaging since the end of the last century [10]. The application of purely air-coupled transducers has always been limited for the detection of nonmetallic materials [11,12] due to the high acoustic impedance between air and solid materials. Other methods, such as disc sensors, have also been widely used for the detection of metallic materials [13], but those methods are not truly noncontact methods. The research team of

Rymantas J. K. developed PMN-PT synthesized with lead magnesium niobate–lead titanate crystals, and applied PMN-PT, which has a strong piezoelectric effect, to an air-coupled transducer to improve performance [14,15]. Therefore, it is believed that the hybrid application of a pulse laser and air-coupled transducer will have good performance as a fully noncontact detection system [16,17].

Most of the works mentioned above are based on the traditional filtered back projection (FBP) algorithm. The FBP algorithm is mainly used for radiographic inspection using X-ray [18,19]. Although high resolution is shown in the radiographic inspection, there is a disadvantage that image artifacts are generated when ultrasonic tests are applied. Other works have been based on the reconstruction algorithm for the probabilistic inspection of damage (RAPID) [20–23]. The RAPID algorithm is actively used in ultrasonic tomographic techniques as a technique for imaging defects using a signal difference coefficient. However, the RAPID algorithm has the advantage of being able to visualize faults with small transmission and reception points, but due to some negative influences such as shape factor (factor of influence), there is a limit to quantifying the defects. It is also not suitable for multidefect imaging.

In this study, in order to overcome the limitations of the resolution of the RAPID algorithm, which is typical of the existing ultrasonic tomographic algorithm, the FBP algorithm, an imaging technique used in the radiography technique, was applied to the ultrasonic technique. We confirmed the problem with the image artifacts generated in the FPB algorithm using ultrasonic waves. Difference of the projection data, Hilbert transform of the difference data, distance weight, and back projection were applied to the FBP algorithm to reconstruct the Difference Hilbert Back Projection (DHB) algorithm. The radiographic imaging techniques require many data points and reliable data for high-resolution imaging. However, manual operation is still being widely used for data acquisition. The work efficiency of manual operation is very low, and the precision of signal acquisition cannot be guaranteed. Although automatic control was used in some studies, many of the devices can only be used in some special circumstances. In this work, a laser switch has been adopted in an automatic detection system to improve the precision of signal acquisition. The automatic detection system was developed based on the hybrid application of a pulse laser and an air-coupled transducer. This is one of the best combinations among many different kinds of noncontact NDT methods. The detection system consists of two main parts. One part is an automatic scanning robot arm for the air-coupled transducer. Another part is an automatic clamping device for different kinds of specimens to be tested. The tomography imaging can be conducted efficiently by using this automatic detection system. High-precision signal acquisition can be guaranteed with the help of the laser switch. The iterative reconstruction method used in this work is the fan-beam method. The practicability of this detection system has been confirmed by experimental results. The experiment results matched well with both the theoretical simulation and the actual defects. The imaging quality has been improved to a great extent compared with other methods used in earlier studies.

2. Tomography Theorem

Tomography imaging was used to obtain the section view of the specimen under testing. Image reconstruction from the data projection is a mathematical problem. The original object can be copied from a large number of projection signals. The results of computed tomography are the distribution image of the linear attenuation coefficients.

The DHB algorithm used in this work has evolved from the traditional FBP algorithm. The experiment results show that the imaging quality obtained from the DHB algorithm is much better than that obtained from other analogous algorithms such as RAPID or FBP. These two frequently used algorithms are also discussed to emphasize the advantages of the new DHB algorithm. The theoretical foundation and the corresponding experimental results will be illustrated in the following part of this paper.

The RAPID is based on a damage index known as the signal difference coefficient (SDC). This SDC can be taken as a measurement of how statistically different the defect signals are from a baseline nondefect signal. Damage indexes for all transducer pairs are spatially distributed. Those damage indexes are summed to generate an image of the damage area. The first step of this algorithm is to calculate SDC between the current signal $x_{ij}(t)$ and the reference $y_{ij}(t)$. The index i is for the transmitter and j is for the receiver. Mathematically, the SDC can be stated as:

$$SDC_{ij} = 1 - \left| \frac{\int_{t_0}^{t_0+\Delta T} [x_{ij}(t) - \mu_x][y_{ij}(t) - \mu_y]^2 dt}{\sqrt{\int_{t_0}^{t_0+\Delta T} [x_{ij}(t) - \mu_x]^2 dt \int_{t_0}^{t_0+\Delta T} [y_{ij}(t) - \mu_y]^2 dt}} \right|, \tag{1}$$

where t_0 is the direct arrival time for each transducer pair, μ_x and μ_y are the mean values of the corresponding current signal and baseline signal, and ΔT is the time window. The next step after the SDC values for all sensor pairs are calculated is image reconstruction. An image is generated by distributing each SDC value spatially on the image plane in an elliptical pattern. The two foci of the ellipses are the two transducers' locations. A parameter β controls the size of the ellipse. The amplitude linearly tapers from its maximum value along the line connecting the ellipse foci to zero on the periphery of the ellipse. The parameter β is a shape factor that controls the size of the elliptical distribution. Its value can be arbitrary, but greater than 1.0. Finally, the image amplitude at each pixel is a linear summation of the location probabilities from each transmitter–receiver pair (M). In Equation (2), the coordinate position of the image is displayed. x_{RK}, y_{RK} is the coordinate position of the sensor with x_{RK}, y_{RK} is the coordinate position of the receiving sensor. $P_K(x, y)$ is the image value of image obtained from a pair of sensors and sensors. $P(x, y)$ is the sum of the image values obtained from the sensors of all pairs (M) and the receiving sensor. $R(x, y, x_{TK}, y_{TK}, x_{RK}, y_{RK})$ is a formula showing the relationship between the coordinate position of the image and the position of the sensor. The larger the $R(x, y, x_{TK}, y_{TK}, x_{RK}, y_{RK})$, the farther away from the shortest distance of the pair of sensors. β is a shape factor that limits the coverage of $R(x, y, x_{TK}, y_{TK}, x_{RK}, y_{RK})$.

$$P(x, y) = \sum_{i=1}^{N-1} P_k(x, y) = \sum_{K=1}^{M} \frac{SDC}{\beta - 1} \{\beta - R(x, y, x_{TK}, y_{TK}, x_{RK}, y_{RK})\} \tag{2}$$

The RAPID has been widely used in the field of nondestructive testing. However, some disadvantages of this algorithm are gradually being exposed with the increasing demands of imaging quality. Those disadvantages are listed as follows. In signal extraction, the time window ΔT in Equation (1) should cover at least two modes of Lamb waves because it is necessary to know the difference between the current signal and baseline signal. A smaller correlation coefficient value means a stronger indication of the defect's appearance. This will undoubtedly make the guided wave tomographic imaging become more complex. Particularly, this adverse effect will become more serious when the air-coupled transducer is used because air-coupled transducers are insensitive to the symmetric modes of a Lamb wave. On the other hand, only one mode of the Lamb wave is needed if the DHB algorithm is used. In defect quantitative analysis, the parameter β determines the shape of the elliptical path. Then, the speckle of the defect shape can also be influenced. There is one optimal value for specific defects with a determined size. Other defects of different sizes cannot be imaged successfully. The influence of β on the imaging results is shown in Figure 1.

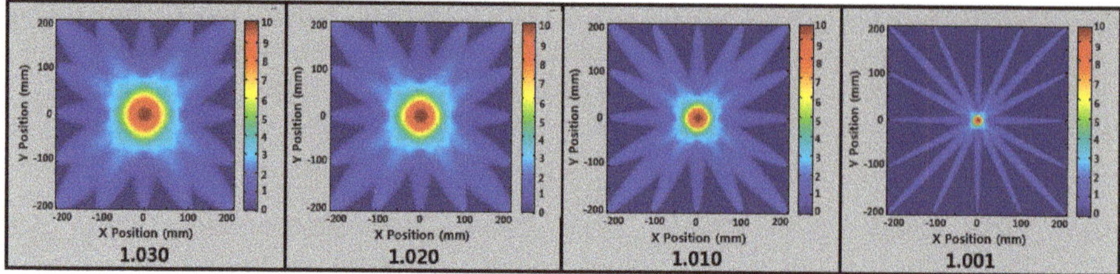

Figure 1. Image reconstruction results by using different values of parameter β.

The FBP algorithm is also widely used in the NDT field. Its reconstruction formula can be written as:

$$f(\vec{x}) = \int_0^{2\pi} \frac{D}{(D+\vec{x}\cdot\vec{e}_w)^2} \int_{-\gamma_m}^{\gamma_m} (q(\lambda,\gamma)\cos\gamma) \otimes h_H(\gamma) d\gamma d\lambda, \tag{3}$$

where $\vec{x}$ is the vector of the reconstruction points, λ is the angle between the starting coordinate axis and the line that connects the trigger points and the origin of the starting coordinate system, λ_m is the angle range of the fan beam, $q(\lambda,\gamma)$ is the projection value, $\vec{e}_w$ is unit orthogonal vector, h_H is kernel of the Hilbert transform, and D is the distance between the trigger points and the intermediate point of the receiving sensors' array.

Many different kinds of iterative reconstruction methods can be used for the data acquisition, as shown in Figure 2. Only the fan-beam method was used in this work to emphasize the comparison of the imaging quality of different tomographic imaging algorithms.

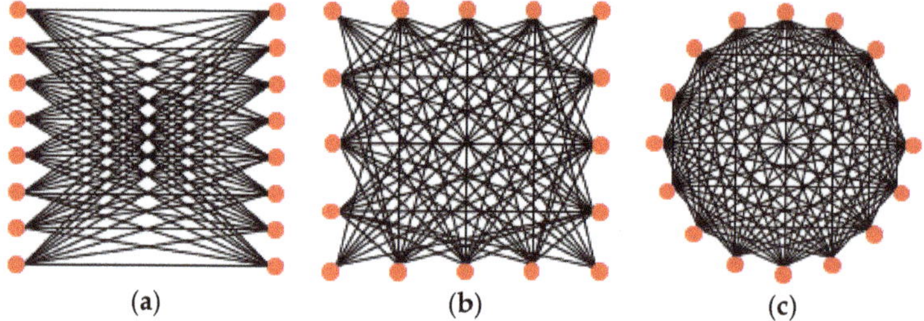

Figure 2. Three common iterative reconstruction methods: (**a**) Crosshole; (**b**) Double-crosshole; (**c**) Fan-beam.

As shown in Figure 3, it is based on the signal of a nondefective ultrasonic signals, and the nondefective signal becomes a reference signal. When the ultrasonic wave passes through the defect, the ultrasonic amplitude is attenuated, so the defect or object is imaged by utilizing the difference in amplitude between the two signals. The simulated results were made based on the attenuation law of a Lamb wave. As shown in Figure 4, results based on different detection points were made for comparison. The FBP algorithm is a technique used in radiographic inspection. It is a method of transmitting X-rays through an object and receiving it through a detector. It shoots 360 degrees around the object and displays the image. However, radiographic inspection is limited by shielding facilities or radiation regulations. In this study, ultrasonic technology using a Lamb wave was applied to compensate for such shortcomings. Figure 4 is the result of imaging by applying the simulation signal of Figure 3 to the FBP algorithm. The imaging quality can be improved by increasing the emitter and receive detection points. This method has the same characteristic as the new DHB algorithm in this respect. The number of the detection points adopted in the actual experiment was 90, 120, and 180.

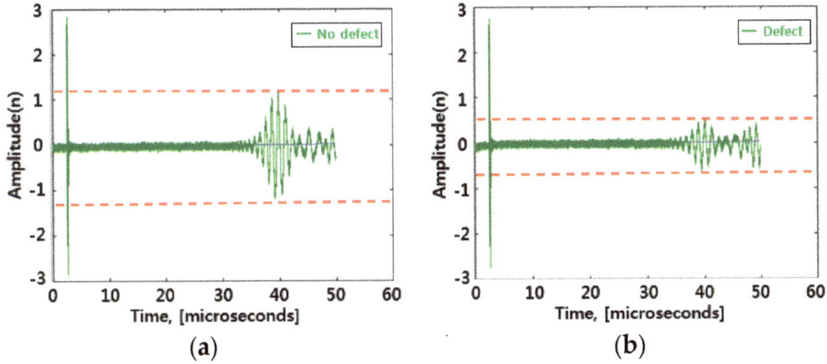

Figure 3. Guided wave signal: (**a**) No defect signal; (**b**) Defect signal.

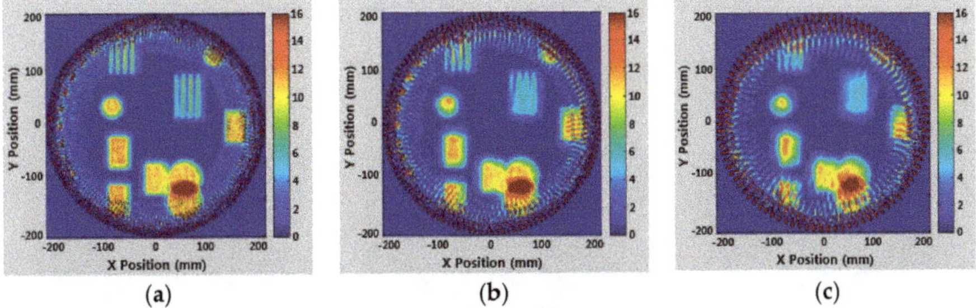

Figure 4. Simulated results from the filtered back projection (FBP) algorithm: (**a**) 180 Points; (**b**) 120 Points; (**c**) 90 Points.

The FBP algorithm has a big advantage in the aspect of imaging precision compared with RAPID. However, the problem of this algorithm is in its image artifacts. Defect imaging close to the scanning edge cannot be displayed clearly, as shown in Figure 4. These are called image artifacts. The image artifacts will undoubtedly affect the precise identification of the defects near the detection boundary.

The new DHB algorithm is proposed to solve the shortcomings of the algorithms discussed above. The new DHB algorithm has good performance in the detection of plate-like structures. The imaging quality can be significantly improved compared with other algorithms, such as RAPID or FBP. Particularly, it can give accurate size and shape of different kinds of defects.

The geometry of the corresponding fan-beam iterative reconstruction method is shown in Figure 5. Its reconstruction formula is Equation (4) [24]. The relevant parameters in Equation (4) are the same as in Equation (3). The unit orthogonal vector $\vec{e}_w$ is represented by $\vec{a}(\lambda)/\|\vec{a}(\lambda)\|$ and is substituted into Equation (3). The notation $g_m(\lambda,\gamma)$ is used to denote the measured projections. Three steps are included in using this method: the difference of the projection data, the Hilbert transform of the difference data, the distance weight, and back projection. The corresponding geometry schematic can be referenced in Figure 5.

$$f(\vec{x}) = \frac{1}{4\pi}\int_0^{2\pi} d\lambda \frac{D}{\|\vec{x}+\vec{a}(\lambda)\|}\int_{-\gamma_m}^{\gamma_m} d\gamma h_H \sin(\gamma'-\gamma)\left(\frac{\partial}{\partial\lambda}+\frac{\partial}{\partial\gamma}\right)g_m(\lambda,\gamma). \quad (4)$$

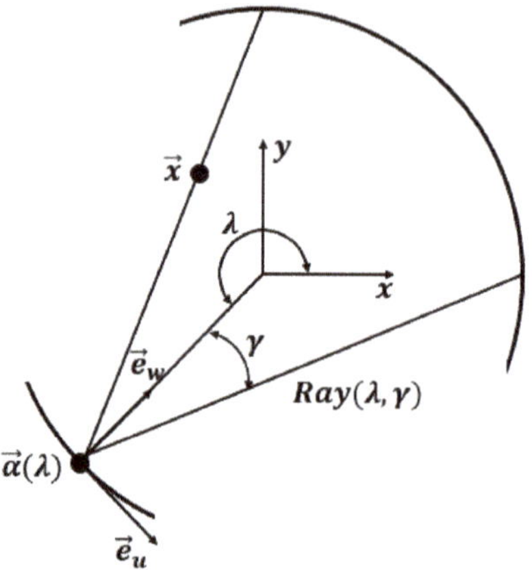

Figure 5. Fan-beam geometry based on the equiangular rays.

The same simulated results have been made by using the DHB algorithm. As shown in Figure 6, the problem of image artifacts can be solved perfectly. The imaging quality is also improved to a large extent.

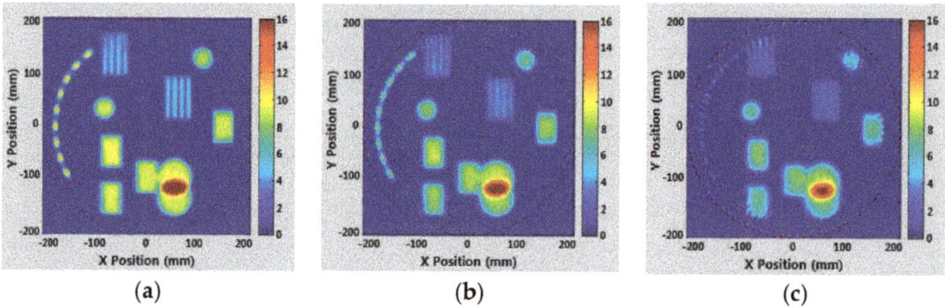

Figure 6. Simulated result from the Difference Hilbert Back Projection (DHB) algorithm: (**a**) 180 Point; (**b**) 120 Point; (**c**) 90 Point.

In addition, short-scan conditions can also be relaxed once the new DHB algorithm is used. It is generally believed that accurate and stable reconstruction of any region of interest (ROI) requires line integrals of the object density for all lines passing through the object. This condition leads to the notion of circular short scan. As shown in Figure 7, $\pi + \theta$ is referred to as a short scan.

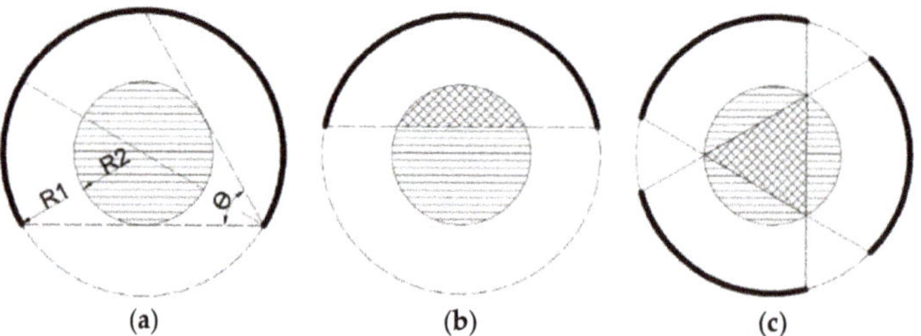

Figure 7. Schematic of the short-scan problem: (**a**) A conventional short-scan of $\pi + \theta$ allows reconstruction of the whole object; (**b**) A continuous scan of less than $\pi + \theta$ allows reconstruction of all object points inside the convex hull of the scan; (**c**) A scan of three equally spaced segments of 80° each allows reconstruction of a triangular ROI in the center of the object.

All line integral measurements through the object are required for any ROI reconstruction using the conventional FBP algorithm. However, this problem can be relaxed using the new DHB algorithm. As shown in Figure 7c, small central ROIs can be reconstructed using several short arcs if only an ROI inside the object needs to be reconstructed. Reconstruction is possible for any ROI inside the central triangular region obtained by joining the beginning point of each arc with the end point of the next arc.

3. Experimental Set-Up and Specimen

The hybrid application of an air-coupled transducer and pulse laser is used in this work. Air-coupled sensors can have increased sensitivities compared to optical transducers. Well-designed capacitive air transducers can be expected to be as sensitive as laser interferometric sensors [25,26]. In addition, the air-coupled transducers can receive a specific mode of guided waves. This can be realized by changing the receiving angle to a leak direction of the selected modes. It is well-known that a static scanning process of the receiver is very important for high-precision acquisition of the signals. The application of air-coupled transducers is one of the most appropriate choices to realize high-precision acquisition. However, the mismatch of acoustic impedances between solid materials and air is a huge weakness of a purely air-coupled transducer detection system. Consider the scenario shown in Figure 8. The ultrasound passes through the interface between the solid specimen and air two times. Only a tiny fraction (10^{-8}) of the original acoustic energy is transmitted through both solid–air interfaces to reach the receiver.

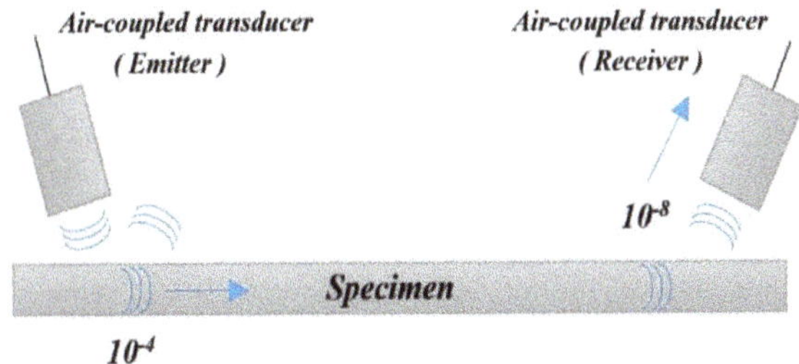

Figure 8. Transmission in air using two air-coupled transducers. Values shown are approximate fractions of the original energy.

Consider the second scenario shown in Figure 9. The ultrasound is generated directly in the solid using a pulsed laser. The solid–air interface on the generation part has effectively been removed. The energy reaching the receiver could increase by a factor of 10^4 [27]. As a receiver, the air-coupled transducer can also have good performance if the proper parameters are set [28]. The hybrid application of an air-coupled transducer and pulse laser can make material characterization more feasible. It can also be used where the air-coupled transducer and pulse laser are put on the opposite sides of the specimen [29,30]. The laser ultrasonic technique can also be used in the detection of a curved surface specimen [31,32].

A line laser source is much better than a point source for long-range detection [33,34]. Many researchers used spatial array illumination sources produced by several means. Those methods include slit masks, lenticular arrays, optical diffraction gratings, multiple lasers, and interference patterns. The linear slit array mask was used in this study, as shown in Figure 10. A Lamb wave of specific modes can be generated selectively by adjusting the line spacing. The distance between two adjacent line laser sources is defined as the line spacing or element gap marked as ΔS. The value of the line spacing is equal to the wavelength of the generated Lamb wave. The difference between it and the element width equals the width of the single line laser source. The width of the single line light source is also known as the line width marked as w. The value of $w/\Delta s$ is known as the duty ratio. It is found that an A0 mode Lamb wave has the optimum signal–noise ratio when the value of $w/\Delta s$ is 0.5 [35].

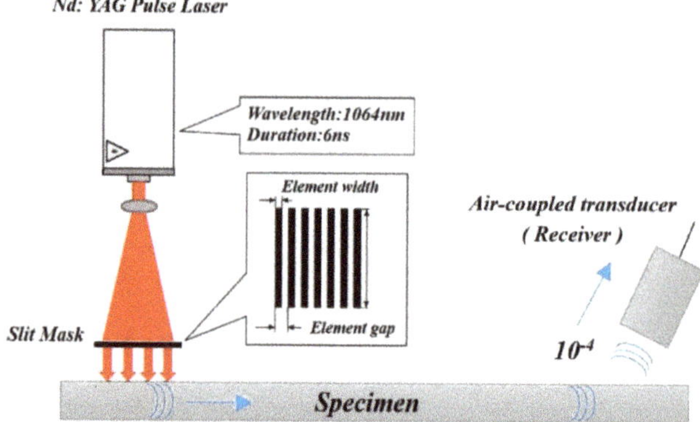

Figure 9. Improved transmission using a laser source. Values shown are approximate fractions of the original energy.

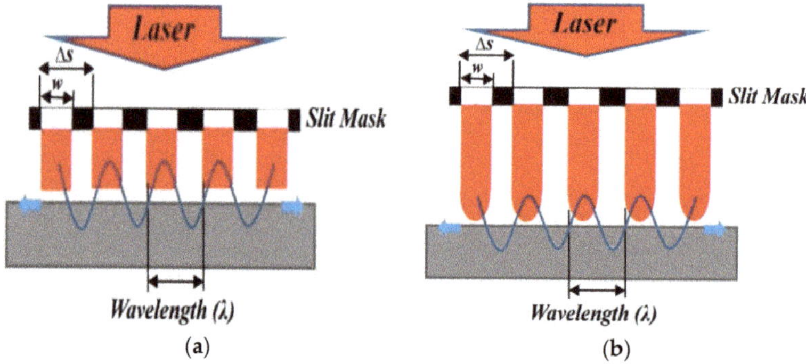

Figure 10. Schematic of the line laser sources of 1-D and Gaussian model: (**a**) One-dimensional theory; (**b**) Gaussian model.

As shown in Figure 10b, the generated wave should be analyzed by using a Gaussian model. Specifically, the spatial energy profile of each line beam is assumed as Gaussian. A Gaussian model is also suitable for the focused laser beams. However, the diffraction effect is negligible when the slit mask is very close to the target surface. Then, the spatial energy profile of each line beam is closer to a square pulse than Gaussian in the line-arrayed slit mask shown in Figure 10a. The energy of each line laser beam can be assumed to have a uniform distribution, and the temporal profile is a delta function. The spatial energy profile of each line beam can be taken as a simple one-dimensional square pulse. For an array of N equally spaced identical laser sources, the final displacement can be taken as a summation of N single-source surface waves displaced in time by the surface wave propagation interval between two adjacent sources, as shown in Equation (5).

$$g(t) = \sum_{n=1}^{N} f(t - n\Delta t) \quad (5)$$

In the above equation, $\Delta t = \Delta s/c$ and $f(t)$ is the out-of-plane displacement of the guided wave generated by a single line laser source. It is expressed as a function of time in the thermoelastic regime. The displacement is proportional to the derivative of the absorbed laser energy function. The energy distribution of the line laser beams was regarded as a square pulse, and $f(t)$ can be approximated by the square pulse derivative. The relationship between the Lamb wave frequency dispersion curve and the corresponding parameters is given in Figure 11.

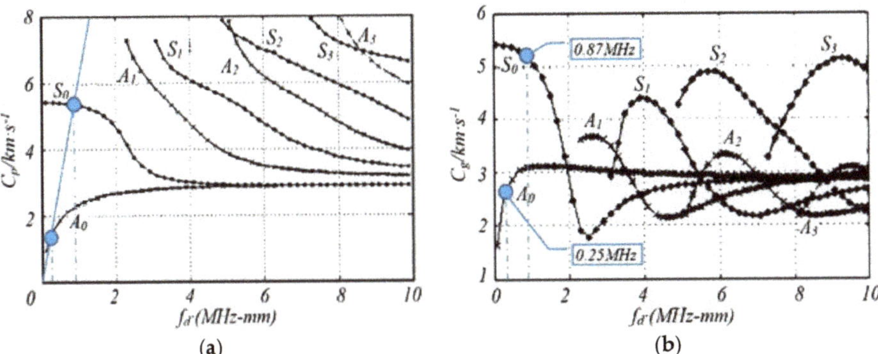

Figure 11. The frequency dispersion curve of a Lamb wave and the corresponding parameters: (**a**) Phased velocity; (**b**) Group velocity.

The signal is generated on the specimen surface when the illuminated laser beam transmits through the slit mask. The wavelength corresponds to the element gap of the slit array. The relation between the wavelength, phase velocity, and wave frequency is shown in Equation (6), where d is the specimen thickness, f is frequency, and λ is wavelength. The principle of generating the desired mode is shown in Figure 11.

$$C_{ph} = f{\cdot}\lambda = f{\cdot}\Delta s = f{\cdot}d{\cdot}\left(\frac{\Delta s}{d}\right). \tag{6}$$

As shown in Figure 11, a diagonal line with a slope of $\Delta s/d$ is described in the dispersion curves of the Lamb wave. The modes of the curves that have cross points with the diagonal line can be generated. Therefore, various modes with different frequencies can be generated simultaneously.

The experimental setup is shown in Figure 12. A Nd:YAG pulsed laser with a wavelength of 1064 nm and a pulse duration of 6 ns was used as the energy source. A Quantel Brilliant B pulsed YAG laser was used with principal radiation at 1064 nm, max energy/pulse 850 mJ, max average power 7.5 W, and pulse duration 6 ns. This laser beam was expanded and illuminated on a slit mask. The slit mask is located close to the specimen surface to suppress the diffraction phenomena.

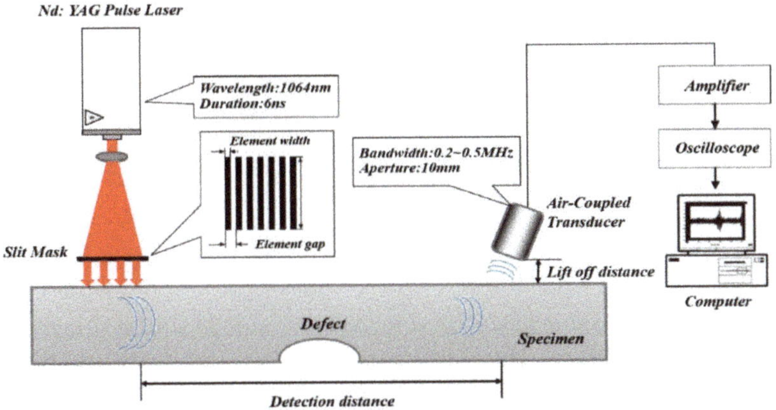

Figure 12. Schematic diagram of the experimental setup.

An unfocused planar air-coupled transducer with a circular aperture was used as the receiver. It was placed on an automatic scanning robot arm. The diameter of its aperture area is 10 mm. The air-coupled transducer used in the experiment is a transducer capable of transmitting and receiving broadcast frequencies. It can detect an ultrasonic wave in air with a frequency range of 0.2 to 0.5 MHz. The lift-off distance of this air-coupled transducer from the specimen surface is 7 mm. The air-coupled transducer has a small amplitude because it receives ultrasonic waves through the air noncontacting the specimen. Therefore, an amplifier is used to amplify the signal. There are two key points that contribute to the imaging quality of the tomography. One is the algorithm, and another one is the high-accuracy acquisition of signals. By using the automatic scanning robot arm, the air-coupled transducer can be positioned automatically with the right receiving angle. The details of the automatic position will be illustrated. The wave propagation direction and the receiving direction of the air-coupled transducer should be kept in the same straight line in practical application. Although the line source is based on a one-dimensional model, the pulse laser has a Gaussian model [36]. This makes the accurate positioning of the air-coupled transducer very important. Thus, an automatic positioning robot arm was developed in this work.

Figure 13 shows the characteristics of the Gaussian beam. In the wave traveling toward the focus $(z = 0)$, the beam width is continuously reduced, so that the beam width is minimized at the focus. Wave past focus increases beam width almost linearly. Suppose that the electromagnetic wave is traveling in the +z direction, and also assume that the Gaussian beam changes like a Gaussian function in the x and y axis directions perpendicular to the z axis. Then, Equation (7) can be obtained.

$$E(\bar{r}) = A(z)\exp\left[i\frac{k\rho^2}{2q(z)}\right]\cdot e^{ikz}, \, where \; \rho^2 = x^2 + y^2 \tag{7}$$

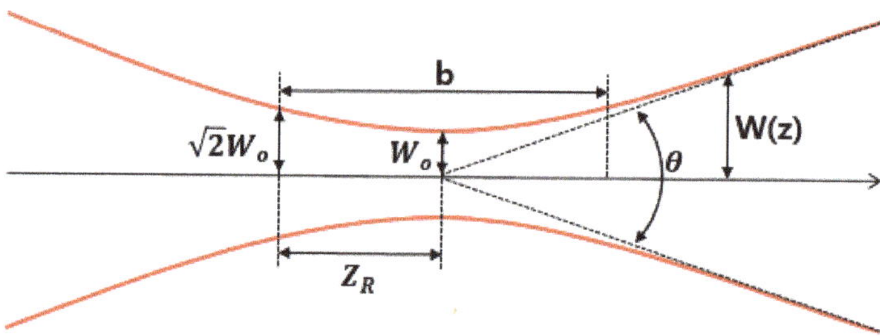

Figure 13. Gaussian beam propagation factor.

Here, K is the wave number, and the gauss beam moves in the +z direction, so the beam width widens and the phase changes, so if $q(z)$ in Equation (8) is defined as plural, Z_R can be defined by considering Figure 13. Referring to Figure 13, W_o has the smallest beam width $(z = 0)$, so measurement is possible. That is, since Wo is half the beam width, Z_R is also determined by $W(z)$. Z_R is called Rayleigh length, and $W(z)$ is increased by $\sqrt{2}$ times at $Z = Z_R$, and the position of $Z = Z_R$, which is doubled by the area of the beam, becomes Rayleigh length.

$$|A(z)| = \sqrt{2\eta P_0}\sqrt{\frac{2}{\pi}}\frac{1}{w(z)} \tag{8}$$

Even if the Gaussian beam travels in the +z direction, the total power must not change, so $A(z)$ can be expressed as Equation (8). Substituting 2 into Equation (7), the Gaussian beam formula is defined as Equation (9). By installing the lens using the formula, a plane wave is generated through the lens, and when passing through the slit, a desired wave can be created.

$$E(\bar{r}) = E_0\sqrt{\frac{2}{\pi}}\frac{w_0}{w(z)}\exp\left[-\frac{\rho^2}{w^2(z)} + i\frac{k\rho^2}{2R(z)}\right]\cdot e^{ikz} \tag{9}$$

As shown in Figure 14, the scanning process consisted of four steps, and the initial condition is given in step 1. The air-coupled transducer should firstly revolve at a certain angle α around the test center when it is necessary to change the detection points. Then, the slit mask is rotated to make the wave propagation direction stay in the same straight line with the new receiving angle. This process can be very complicated with manual computation and operation. Thus, automatic control was adopted in this work. The LED laser line sources and photoresistors were installed on the edge of the slit mask and the rotation center of the air-coupled transducer. They were marked as $LED_{1/2}$ and $R_{1/2}$, respectively. All the automatic movement in this work is based on a step motor. To rotate the slit mask, LED_1 will be turned on at the same time as the step motor M_1. The illumination direction of LED_1 is the same as that of the wave propagation. The step motor M_1 will stop by itself once the light of LED_1 points to the photoresistor R_1. Then, LED_2 will turn on together with M_2, and the step motor M_2 will also stop automatically once the light of LED_2 points to the photoresistor R_2. The new detection point can be fixed with the right receiving angle automatically based on the process discussed above. The complex calculations and manual operation can be omitted.

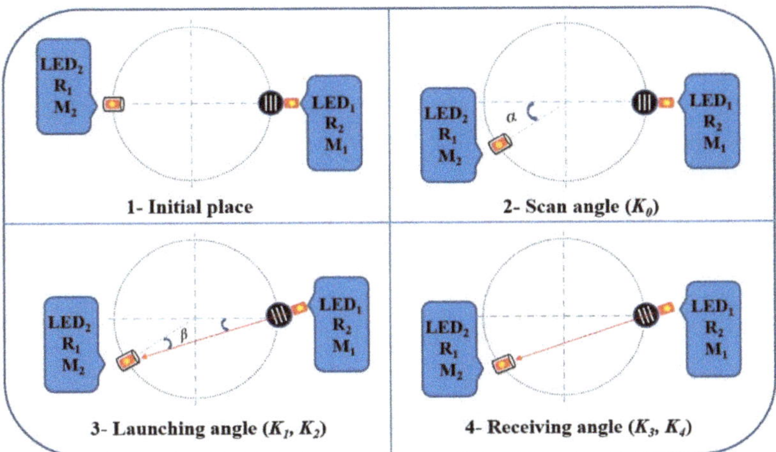

Figure 14. Schematic of the automatic scanning robot arm.

The design of the automatic scanning is shown in Figure 15. The key part here is the photoresistor. Its electrical resistance will decrease immediately if it is illuminated by the light of the laser LED. Then, the corresponding circuit that was used to control the step motor will close. Conversely, the circuit will turn off. In this work, a single chip computer was used to control the step motor. The step motor will be fixed by its magnetic pole once the circuit is controlled by the photoresistor. It can guarantee that the experimental equipment will never be influenced by potential factors from outside. In other words, the movement of the transducer can only be controlled by a program. Then, the detection points can be positioned with very high accuracy with the right receiving angle of the air-coupled transducer.

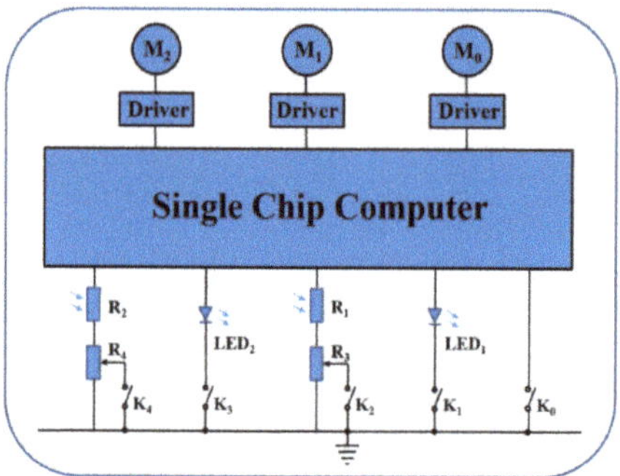

Figure 15. The working principle of the automatic scanning robot arm.

In a practical environment, the photoresistor can also be influenced by sunlight or the bulbs in the lab. Those negative factors can influence the normal operation of the photoresistor, as shown in Figure 16. Adjustable resistors R_3 and R_4 were installed in series with the photoresistors to avoid these influences. They can improve the adaptability of the equipment in different working environments. The photoresistor will only respond to the laser LED by using the method mentioned above. Both the air-coupled transducer and the pulse laser are on the same side of the specimen in the experimental setup. The automatic clamping device will be put on the other side of the specimen. The accurate collection of the signals cannot only rely on the automatic positioning of the air-coupled transducer. The precise movement and fixation of the specimen are equally as important as they are for the air-coupled transducer. Accurate collection can be achieved by using an automatic clamping robot. As shown in Figure 17, it can clamp different kinds of specimens automatically by using four robot arms. The specimens include plates, curved plates, or even pipes. There are four lead-screw mechanisms that are used to control the robot arms. Those robot arms can stop by themselves once they come into contact with the specimen boundaries. The movement of those robot arms will never be influenced due to the self-locking phenomenon of the lead-screw mechanisms.

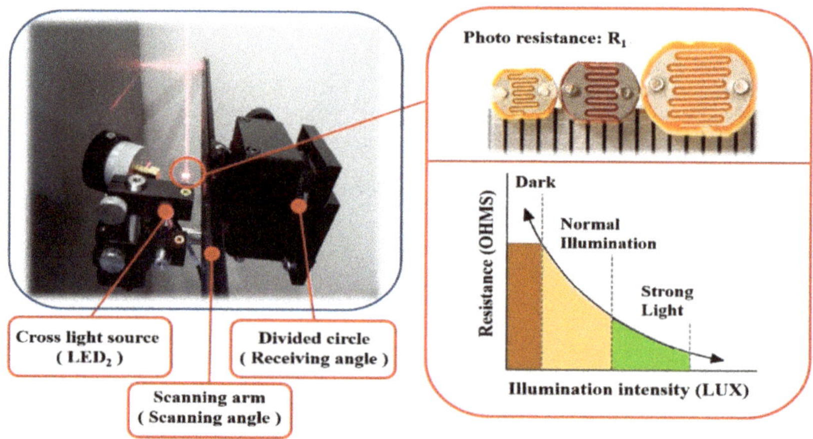

Figure 16. The actual settings of the automatic scanning robot arm.

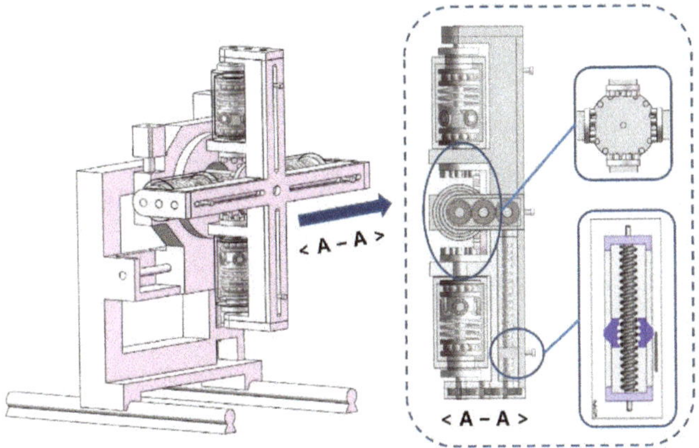

Figure 17. The 3-D model of the automatic clamping robot arms.

The key part of the automatic clamping robot arm is an automatic clutch. The robot arm can clamp the plate specimen with different shapes automatically by using this part, and the clamping force is adjustable. As shown in Figure 18, two transmission shafts are connected together based on elastic force. Those two shafts will always be connected if F_p is not bigger than a maximum value. Otherwise, no force can be transferred through this mechanism. The corresponding calculation formulas are shown in Equations (10) and (11).

$$F_p = F_e < F'_e \tag{10}$$

$$F'_p = F'_e = -k \cdot x \tag{11}$$

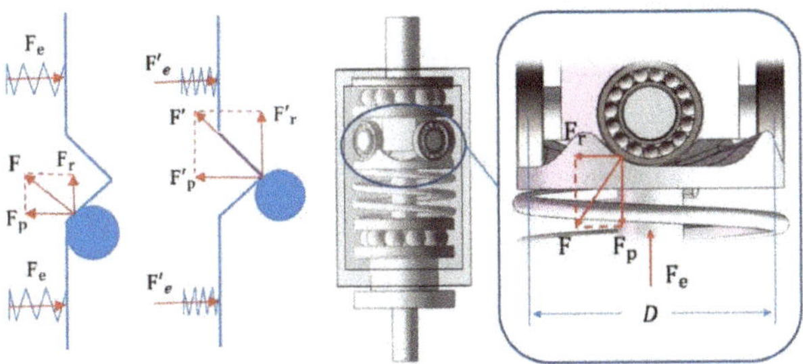

Figure 18. The force analysis of the automatic clutch.

Each of the four robot arms can stop by itself if it comes into contact with the specimen's boundaries. At the same time, the other robot arms will continue to clamp the specimen until they also come into contact with the specimen's boundaries.

In this work, all of the linear motion is based on the lead screw mechanisms, and all of the circular motion is based on worm and gear mechanisms. A self-locking effect can be realized by using both mechanisms. The reliability of the necessary movement in actual operation can be guaranteed in a further step based on the self-locking effect. The actual experimental platform is shown in Figure 19. The air-coupled transducer was placed on the same side of the specimen together with the pulse laser. The clamping robot arms was fixed on the other side of the specimen. The clamping robot arms were installed on a scanning device through which high-accuracy positioning of the specimen can be achieved.

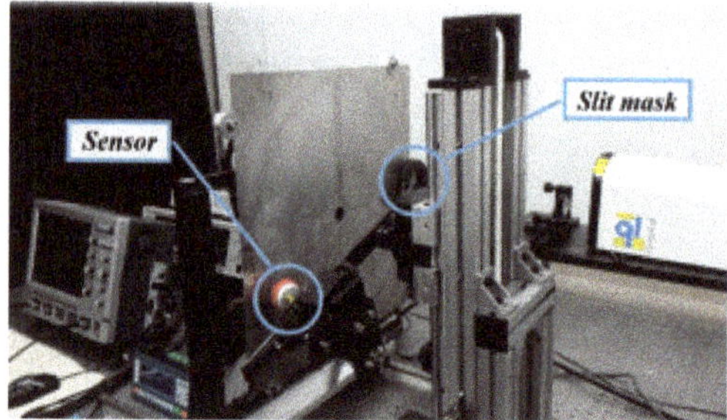

Figure 19. The actual experimental platform.

4. Results and Discussion

In this work, the test specimen is an aluminum plate with a thickness of 1 mm. The wavelength was designed as 6 mm. Both the A_0 and S_0 modes of a Lamb wave can be excited based on the laser ultrasonic technique. Because the air-coupled transducer is more sensitive to the antisymmetric modes than the symmetric modes, the detection range of the air-coupled transducer was chosen as 0.2 to 0.5 MHz to detect the A_0 mode Lamb wave.

An experiment was done to see the appearance of a defect, as shown in Figure 20. There is a penetrating hole in the center of the specimen. The diameter of this through hole is 20 mm.

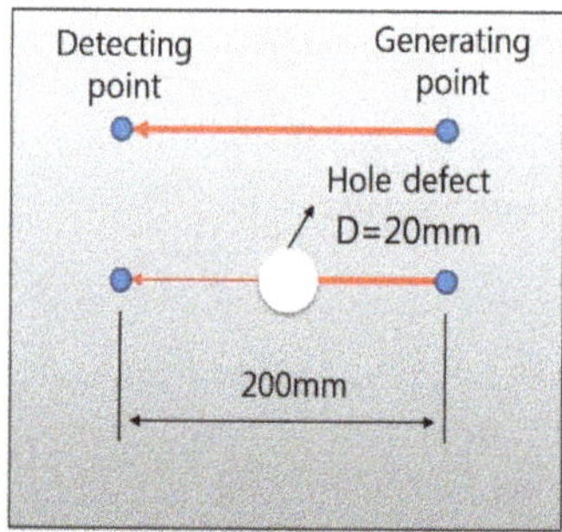

Figure 20. The schematic of the defect-detecting experiment.

The signal of both the defect-free and with the defect areas have been obtained in both time and frequency domains. As shown in Figure 21, the signal attenuates clearly due to the appearance of the defect. Only an A_0 mode Lamb wave can be detected, as mentioned above. The generated S_0 mode was filtered by the air-coupled transducer due to its detection range. Only one mode of the Lamb wave was needed for the tomography once the new DHB method was used. This is also another advantage of the new algorithm. For example, the probabilistic algorithm requires at least two modes of Lamb waves for comparison. The tomographic imaging technique used in this work is based on the wave attenuation. The attenuation law of a Lamb wave is subject to Equation (12), and the equation for data projection is given in Equation (13). The parameter α in those two equations is the attenuation coefficient, A is the signal amplitude, and x is the propagation distance.

$$A = A_0 e^{-\alpha x} \tag{12}$$

$$\int_{-\infty}^{\infty} \alpha(x)dx = \ln \frac{A_0}{A} \tag{13}$$

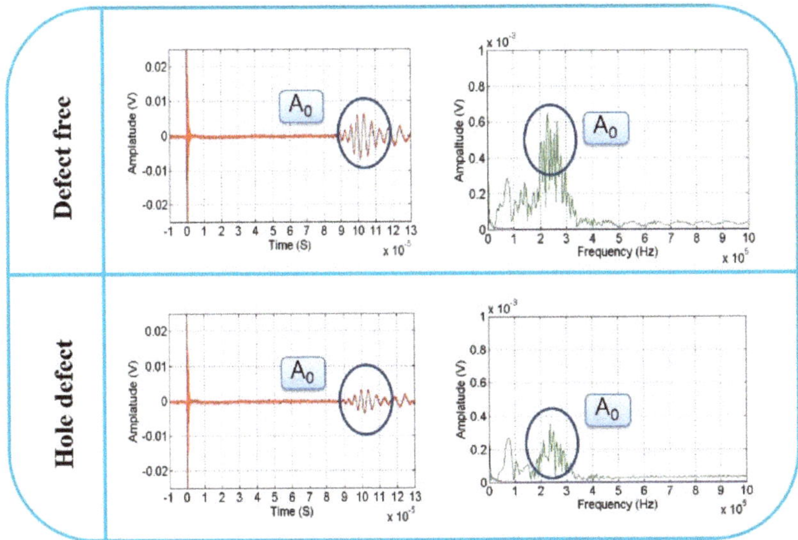

Figure 21. The signals of the defect detecting experiment.

The attenuation curve and the corresponding scanning sketch based on the fan-beam iterative method are shown in Figure 22a,b. The signal attenuation becomes evident within the scanning range of 83° to 90° in the area where a defect starts to appear. Particularly, the illumination area can cause a "dead zone" due to the geometrical features (the diameter of the slit mask used in this work is 60 mm). In other words, the detection cannot be achieved effectively along a full circle arc when the line laser source is used. As shown in Figure 22, the signal within the scanning range of 0° to 13° was invalid data.

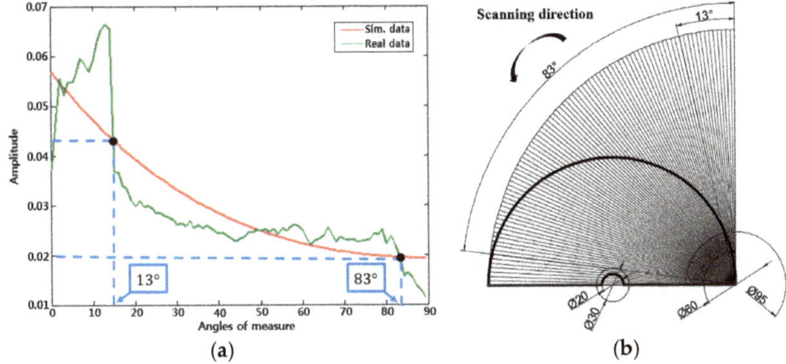

Figure 22. The attenuation curve of Lamb wave obtained from the experiment: (**a**) Attenuation curve; (**b**) Schematic of scanning direction.

It is unnecessary to test the "dead zone" when the DHB algorithm is used because the short-scan condition can be relaxed to a great extent. The real data shown in Figure 22 also supports the advantage of this new DHB algorithm compared with FBP or other algorithms. Figure 23a,b is a sample drawing and specimen of the initial single defect. The imaging of the penetrating hole is given in Figure 24. There are 36 emitter and receive points. If there are 19 receiving points from a single emitter point to a semicircle, and inspection is performed up to the 20th emitter point, 361 data points can be collected.

Complex shapes or multiple defects should be inspected up to the 36th emitter point, but a single circular defect, as shown in Figure 24, was able to quantitatively represent the defect. The imaging of the defect is displayed clearly compared with the defect-free area.

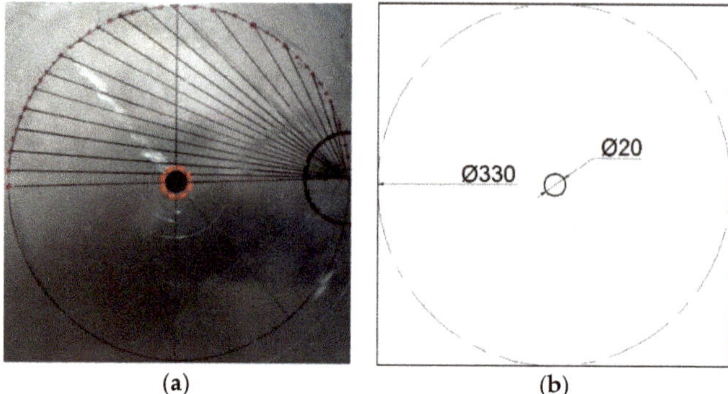

Figure 23. Circular defect specimen #1: (**a**) Circular defect specimen; (**b**) Circular defect specimen drawing (Unit: mm).

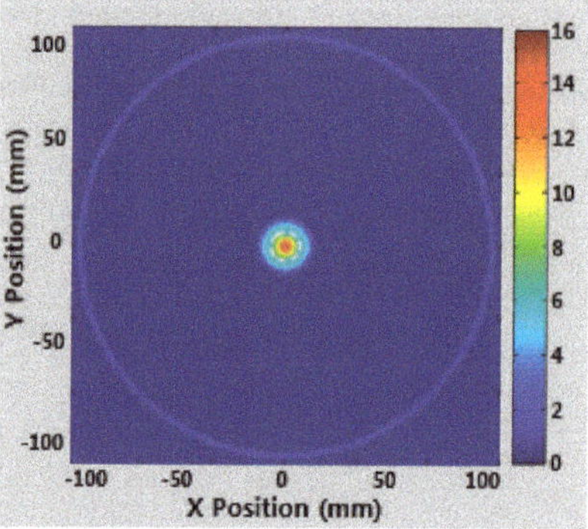

Figure 24. Circular defect imaging.

Another experiment was conducted to further verify the effectiveness of the new DHB algorithm in the aspect of detection precision. Different types of defects were made in different positions of specimen #2. As shown in Figure 25, the defects include 10 micropores, one narrow through defect (width 2 mm, length 52 mm), and two nonpenetrating defects (circle defect: diameter 52 mm, depth 0.5 mm) (rectangular defect: width 52 mm, length 26 mm, depth 0.2 mm) with different depth, which can be considered as an imitation of corrosion defects. There are 106 emitter and receiver points. If there are 53 receiving points from a single emitter point to a semicircle, and inspection is performed up to the 106th emitter point, 5618 data points can be collected. Specimen #2 has multiple defects in various shapes, so the more imitators and reception points, the more quantitatively the defects can be represented. Although it has so many emitters and receiving points, it was able to receive uniform data through the robot arm. Figure 26 shows the imaging results of specimen #2. The imaging results agree very well with the actual defects.

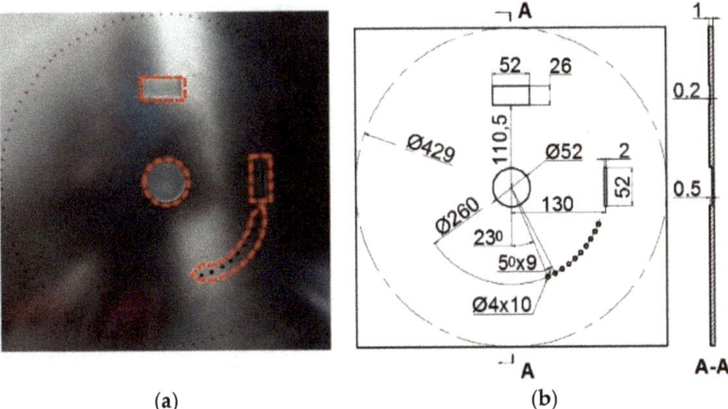

(a) (b)

Figure 25. Image reconstruction of specimen #2: (**a**) Multidefect specimen; (**b**) Multidefect specimen drawing (Unit: mm).

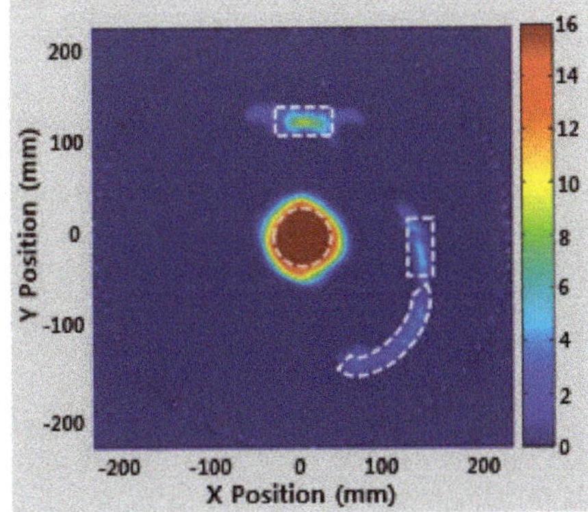

Figure 26. Multidefect imaging using the DHB algorithm.

5. Conclusions

The hybrid application of an air-coupled transducer and pulse laser has been adopted in this paper. The FBP algorithm, which is mainly used in radiographic inspection, was applied to ultrasonic testing and tomographic imaging, and the disadvantages of image artifacts were confirmed. The factors affecting image quality, such as image artifacts, were overcome with the DHB algorithm through three projection data processing steps; Hilbert transform of the difference data, distance weight, and back projection. A new DHB algorithm has been successfully applied to give accurate imaging of different kinds of defects in plate-like structures. The imaging results agree very well with the actual defects. Particularly, the detection efficiency has been improved to a large extent by using the automatic scanning system. Further research will focus on the detection of different kinds of specimens with different geometries, such as curved surfaces. The algorithm should be further improved to make high-quality images with fewer detection points. The presented techniques can be applied to structural health monitoring to predict faults by quantitatively imaging the growth status of the faults.

Author Contributions: J.P. and Z.L. designed and performed the experiments. Y.C. conceived the original idea. J.P. developed the theory and performed the computations with support from J.L. and Y.C. verified the analytical methods. J.L. and Z.L. wrote the draft paper, J.P. completed the final paper, and Y.C. made the final review. All authors have read and agreed to the published version of the manuscript.

Funding: This work was supported by National Research Foundation of Korea (NRF) grant funded by Korea government (MSIT)(NRF-2020M2D2A1A02069933).

Acknowledgments: Authors appreciate the support from Wang Yu at University of Electronic Science and Technology of China.

Conflicts of Interest: The authors declare no conflict of interest.

References

1. Jansen, D.; Hutchins, D.A. Immersion tomography using Rayleigh and Lamb waves. *Ultrasonics* **1992**, *30*, 245–254. [CrossRef]
2. Hutchins, D.A.; Jansen, D.P.; Edwards, C. Non-contact Lamb wave tomography using laser-generated ultrasound. In Proceedings of the IEEE Ultrasonics Symposium, Tucson, AZ, USA, 20–23 October 1992; Volume 2, pp. 883–886.
3. Hutchins, D.A.; Jansen, D.; Edwards, C. Lamb-wave tomography using non-contact transduction. *Ultrasonics* **1993**, *31*, 97–103. [CrossRef]
4. Ho, K.S.; Billson, D.R.; Hutchins, D.A. Ultrasonic Lamb wave tomography using scanned EMATs and wavelet processing. *Nondestruct. Test. Eval.* **2007**, *22*, 19–34. [CrossRef]
5. McKeon, J.C.P.; Hinders, M.K. Parallel projection and crosshole Lamb wave contact scanning tomography. *J. Acoust. Soc. Am.* **1999**, *106*, 2568–2577. [CrossRef]
6. Li, W.; Cho, Y. Combination of nonlinear ultrasonics and guided wave tomography for imaging the micro-defects. *Ultrasonics* **2016**, *65*, 87–95. [CrossRef]
7. Lee, J.; Sheen, B.; Cho, Y. Multi-defect tomographic imaging with a variable shape factor for the RAPID algorithm. *J. Vis.* **2015**, *19*, 393–402. [CrossRef]
8. Lee, J.; Sheen, B.; Cho, Y. Quantitative tomographic visualization for irregular shape defects by guided wave long range inspection. *Int. J. Precis. Eng. Manuf.* **2015**, *16*, 1949–1954. [CrossRef]
9. Park, J.; Cho, Y. A study on guided wave tomographic imaging for defects on a curved structure. *J. Vis.* **2019**, *22*, 1081–1092. [CrossRef]
10. Wright, W.; Hutchins, D.A.; Jansen, D.; Schindel, D. Air-coupled Lamb wave tomography. *IEEE Trans. Ultrason. Ferroelectr. Freq. Control.* **1997**, *44*, 53–59. [CrossRef]
11. Liu, Z.; Yu, H.; Fan, J.; Hu, Y.; Cunfu, H.; Wu, B. Baseline-free delamination inspection in composite plates by synthesizing non-contact air-coupled Lamb wave scan method and virtual time reversal algorithm. *Smart Mater. Struct.* **2015**, *24*, 045014. [CrossRef]
12. Rymantas, J.K.; Almantas, V.; Sustina, S. Application of air-coupled ultrasonic arrays for excitation of a slow antisymmetric lamb wave. *Sensors* **2018**, *18*, 2636.
13. Zhao, X.; Royer, R.L.; Owens, S.E.; Rose, J.L. Ultrasonic Lamb wave tomography in structural health monitoring. *Smart Mater. Struct.* **2011**, *20*, 105002. [CrossRef]
14. Rymantas, J.K.; Reimondas, S.; Justina, S. Application of PMN-32PT piezoelectric crystals for novel air-coupled ultrasonic transducers. *Phys. Procedia* **2015**, *70*, 869–900.
15. Rymantas, J.K.; Reimondas, S.; Justina, S. Air-coupled ultrasonic receivers with high electromechanical coupling PMN-32% PT strip-like piezoelectric elements. *Sensors* **2017**, *17*, 2365.
16. Albiruni, F.; Cho, Y.; Lee, J.-H.; Ahn, B.-Y. Non-Contact Guided Waves Tomographic Imaging of Plate-Like Structures Using a Probabilistic Algorithm. *Mater. Trans.* **2012**, *53*, 330–336. [CrossRef]
17. Park, J.; Lim, J.; Cho, Y. Guided-Wave Tomographic Imaging of Plate Defects by Laser-Based Ultrasonic Techniques. *J. Korean Soc. Nondestruct. Test.* **2014**, *34*, 435–440. [CrossRef]
18. Qi, Z.; Chen, G.-H. Direct Fan-Beam Reconstruction Algorithm via Filtered Backprojection for Differential Phase-Contrast Computed Tomography. *X-ray Opt. Instrum.* **2008**, *2008*, 835172. [CrossRef]
19. Bernal, J.; Sanchez, J. Use of Filtered Back-Projection Methods to Improve CT Image Reconstruction. Available online: http://www.researchagate.net/publication/265811375 (accessed on 4 May 2020).
20. Hay, T.R.; Royer, R.L.; Gao, H.; Zhao, X.; Rose, J.L. A comparison of embedded sensor Lamb wave ultrasonic tomography approaches for material loss detection. *Smart Mater. Struct.* **2006**, *15*, 946–951. [CrossRef]
21. Van Velsor, J.K.; Gao, H.; Rose, J.L. Guided-wave tomographic imaging of defects in pipe using a probabilistic reconstruction algorithm. *Insight Non-Destr. Test. Cond. Monit.* **2007**, *49*, 532–537. [CrossRef]

22. Huthwaite, P.; Simonetti, F. High-resolution guided wave tomography. *Wave Motion* **2013**, *50*, 979–993. [CrossRef]
23. Hettler, J.; Tabatabaeipour, M.; Delrue, S.; Abeele, K.V.D. Linear and nonlinear guided wave imaging of Impact damage in CFRP using a probabilistic approach. *Materials* **2016**, *9*, 901. [CrossRef] [PubMed]
24. Noo, F.; Defrise, M.; Clackdoyle, R.; Kudo, H. Image reconstruction from fan-beam projections on less than a short scan. *Phys. Med. Biol.* **2002**, *47*, 2525–2546. [CrossRef] [PubMed]
25. Boltz, E.S.; Fortunko, C.M. Absolute sensitivity limits of various ultrasonic transducers. In Proceedings of the Ultrasnics Symposium, Seattle, WA, USA, 7–10 November 1995; Volume 2, pp. 951–954.
26. Robert, E.G. Non-contact ultrasonic techniques. *Ultrasonics* **2004**, *42*, 9–16.
27. Wright, W.M.D.; Hutchins, D.A. Air-coupled ultrasonic testing of materials using broadband pulses in through-transmission. *Ultrasonics* **1999**, *37*, 19–22. [CrossRef]
28. Nishino, H.; Masuda, S.; Yoshida, K.; Takahashi, M.; Hoshino, H.; Ogura, Y.; Kitagawa, H.; Kusumoto, J.; Kanaya, A. Theoretical and Experimental Investigations of Transmission Coefficients of Longitudinal Waves through Metal Plates Immersed in Air for Uses of Air Coupled Ultrasounds. *Mater. Trans.* **2008**, *49*, 2861–2867. [CrossRef]
29. Hutchins, D.A.; Wright, W.; Hayward, G.; Gachagan, A. Air-coupled piezoelectric detection of laser-generated ultrasound. *IEEE Trans. Ultrason. Ferroelectr. Freq. Control.* **1994**, *41*, 796–805. [CrossRef]
30. Gachagan, A.; Hayward, G.; Wright, W.M.D.; Hutchins, D.A. Air coupled piezoelectric detection of laser generated ultrasound. In Proceedings of the Ultrasonics Symposium, Baltimore, MD, USA, 31 October–3 November 1993.
31. Yashiro, S.; Takatsubo, J.; Miyauchi, H.; Toyama, N. A novel technique for visualizing ultrasonic waves in general solid media by pulsed laser scan. *NDT E Int.* **2008**, *41*, 137–144. [CrossRef]
32. Song, B.M.; Kim, J.-H.; Lee, J.-H. Application of laser-generated ultrasound to evaluated wall thinned pipe. In Proceedings of the IVth NDT in Progress, Prague, Czech Republic, 5–7 November 2007; pp. 187–195.
33. Achenbach, J.D. Laser excitation of surface wave motion. *J. Mech. Phys. Solids* **2003**, *51*, 1885–1902. [CrossRef]
34. Arias, I.; Achenbach, J.D. Thermoelastic generation of ultrasound by line-focused laser irradiation. *Int. J. Solids Struct.* **2003**, *40*, 6917–6935. [CrossRef]
35. Choi, S.; Jhang, K.-Y. Influence of slit width on harmonic generation in ultrasonic surface waves excited by masking a laser beam with a line arrayed slit. *NDT E Int.* **2013**, *57*, 1–6. [CrossRef]
36. Cosenza, C.; Kenderian, S.; Djordjevic, B.B.; Green, R.E.; Pasta, A. Generation of narrowband antisymmetric lamb waves using a formed laser source in the ablative regime. *IEEE Trans. Ultrason. Ferroelectr. Freq. Control.* **2006**, *54*, 147–156. [CrossRef] [PubMed]

Review

Acoustic Scattering Models from Rough Surfaces: A Brief Review and Recent Advances

Michel Darmon [1,*], Vincent Dorval [1] and François Baqué [2]

1 Université Paris-Saclay, CEA, List, F-91120 Palaiseau, France; vincent.dorval@cea.fr
2 French Alternative Energies and Nuclear Energy Commission—Division of Nuclear Energy CEA-DES, 13108 Saint-Paul-lez-Durance CEDEX, France; francois.baque@cea.fr
* Correspondence: michel.darmon@cea.fr

Received: 22 September 2020; Accepted: 18 November 2020; Published: 23 November 2020

Abstract: This paper proposes a brief review of acoustic wave scattering models from rough surfaces. This review is intended to provide an up-to-date survey of the analytical approximate or semi-analytical methods that are encountered in acoustic scattering from random rough surfaces. Thus, this review focuses only on the scattering of acoustic waves and does not deal with the transmission through a rough interface of waves within a solid material. The main used approximations are classified here into two types: the two historical approximations (Kirchhoff approximation and the perturbation theory) and some sound propagation models more suitable for grazing observation angles on rough surfaces, such as the small slope approximation, the integral equation method and the parabolic equation. The use of the existing approximations in the scientific literature and their validity are highlighted. Rough surfaces with Gaussian height distribution are usually considered in the models hypotheses. Rather few comparisons between models and measurements have been found in the literature. Some new criteria have been recently determined for the validity of the Kirchhoff approximation, which is one of the most used models, owing to its implementation simplicity.

Keywords: acoustic wave; scattering; random rough surface; impedance

---

1. Introduction

The scattering of waves from rough surfaces is an important phenomenon in diverse areas of science, such as optics, geophysics, communications, acoustics, hydraulics [1], elastodynamics for non-destructive testing, etc. For instance, in the particular case of ultrasonic waves, scattering from rough surfaces leads to many applications in non-destructive testing [2], in underwater acoustics for high resolution ultrasonic sonars [3] and for medical sciences [4,5]. The modeling of such phenomena is a century-old and still active research topic.

Scattering models of acoustic waves from rough surfaces could be classified into analytical models, numerical ones and combinations of both analytical and numerical methods. However, such a classification is rather ambiguous, since some analytical models (usually called semi-analytical) employ numerical calculations and since some numerical methods start from analytical approximations. The current review focuses on analytical or semi-analytical methods and the validity of each method will be emphasized as well as some applications examples. Some analytical methods are presented or used in the following in more detail: the Kirchhoff approximation (KA), the perturbation theory (PT), the integral equation (IE) method, the parabolic equation (PE) method and the small slope approximation (SSA). The Kirchhoff and PT methods represent early classical approaches but are still very often employed in the literature, notably the Kirchhoff one. The other methods represent more modern approaches which have larger domains of validity but are more complex and therefore more difficult to implement or can lead to a higher calculation time. All the previously cited methods have been found to be amongst the most common in the literature and many of the other methods of the literature are based on or have much in common with these approaches.

This review focuses thus only on the scattering of acoustic waves and does not deal with the transmission through a rough surface of waves within a solid material. Indeed the models used in acoustics are often also employed in elastodynamics. For instance, the elastic Kirchhoff model has been devised to deal with rough defects [6–8] and to take into account mode-conversions between compressional and shear waves. A variety of analytical models involving reflection and transmission waves (for instance through a rough fluid–solid interface), such as the phase screen approximation [9–12] and Kirchhoff theory, have been developed to understand the elastic fields in solids or scattered from different embedded scatterers.

Several reviews of scattering from rough surfaces exist in the literature, notably in general physics (diverse areas of sciences) for random rough surfaces [13–15] or in electromagnetism [16,17]. The main review describing works in acoustic scattering from rough surfaces has been written by Ogilvy in the 80s [18], although its review does not only deal with acoustics. This review has notably summarized the two classical approximations (Kirchhoff and the perturbation theory) but it is now out of date: since this review, important models have been devised for grazing incidence or observation, such as the small slope approximation, and new criteria have notably been established for defining the validity of the Kirchhoff approximation, which is probably the most used analytical model for roughness scattering. Some other textbooks also deal with acoustic scattering from rough surfaces notably in the view of ocean applications [19,20]. At Ogilvy's time, the accuracy of Kirchhoff approximation came from Thorsos' work [21] and was determined by the ratio of the height correlation length L of the random rough surface to the wavelength. It is also the case in Voronovich's more recent textbook (see Figure 1) [20], where approximations KA, PT and SSA are all said to require this ratio less than a constant in order to be valid; another different criterion has even been proposed for radar applications [22]. Recent studies—described in the present review and based on numerical calculations—have shown for elastodynamic problems that the accuracy of the Kirchhoff theory should depend on the parameter σ^2/L (σ and L being the RMS surface height and the height correlation length). Kirchhoff is considered accurate when this parameter is smaller than an upper bound c depending on the wavelength: ($\sigma^2/L < c$). Kirchhoff accuracy is also dependent on incidence and scattering angles. In the recent years new developments or comparisons of the previous methods have also been proposed in the literature and for different applications, such as non-destructive evaluation (NDE), underwater acoustics or even medicine.

Section 1 of the current paper is devoted to the parametrization of rough surfaces, Gaussian height distribution being usually assumed in modeling. The two historical approximations (Kirchhoff and the perturbation theory) are then described (Section 2) as well as their use in the scientific literature, their validity and an application example for non-destructive testing. Then, more modern approaches with larger domains of validity (notably suitable for grazing observation angles and generally having applications for underwater acoustics) are presented in Section 3.

2. Rough Surfaces Parametrization

This section is devoted to mathematical methods used to characterize randomly rough surfaces. A rough surface is described using its height deviation from a mean smooth surface. Hereafter are presented the statistical distributions (height and correlation functions) and associated parameters usually employed to properly define the rough surface properties. We will here take the example of Gaussian height distributions since most of the used models of rough surface scattering assume such a distribution. However, although many surfaces (both natural and engineered) exhibit Gaussian height distributions, there are also many others that do not [23,24]. Examples of generating non-Gaussian rough surfaces can be found in the literature [25,26].

2.1. Statistical Properties

Surface roughness is described here as the variation of height $z = h(x, y)$ above or below a mean surface. <*h*>, the average of the height variation over the surface, is by definition 0. h is assumed to be random and described statistically using parameters such as the RMS height and spatial correlation lengths. In the theory presented hereafter and developed by Ogilvy [6,27], it is assumed that the height distribution is well represented by a Gaussian.

The probability of any measured height being between h and h + dh above or below the mean plane is then given by p(h) dh, where:

$$p(h) = \frac{1}{\sigma(2\pi)^{\frac{1}{2}}} e^{-h^2/2\sigma^2} \quad (1)$$

where σ is the RMS height:

$$\sigma = \left[\langle h^2 \rangle\right]^{\frac{1}{2}} \quad (2)$$

Random rough surfaces show some spatial correlation, as the height at a location is not totally independent of surrounding heights. This may be described by the correlation function of a surface, $C(\mathbf{R})$ where:

$$C(\mathbf{R}) = \frac{1}{\sigma^2} \langle h(\mathbf{r}) h(\mathbf{r} + \mathbf{R}) \rangle \quad (3)$$

which represents the extent to which the surface height at any one point **r** is, on average, related to the surface height at point **r** + **R** away. Gaussian functions are often used as an approximation of the correlation of real surfaces. Assuming the same Gaussian correlation in every direction, C can be written:

$$C(\mathbf{R}) = e^{-R/L^2} \quad (4)$$

where L is called the correlation length of the surface. Anisotropic correlation functions can also be used to describe surface with a direction-dependent correlation. Such surfaces often occur in practice in industrial structures, as roughness can be generated in anisotropic manner during the manufacturing process.

2.2. Generation Process of an Individual Gaussian Rough Surface

The method of generating these surfaces is described in detail in [6,27]. It first generates uncorrelated random numbers with a Gaussian distribution of zero mean and specified RMS value. Then, it performs a moving average with weights that ensure that the correlation function is Gaussian with the desired correlation length. The following weights yield a surface with two independent lengths L_x and L_y in the two perpendicular directions:

$$W_{ij} = e^{-2[(i\Delta x)^2/L_x^2 + (i\Delta y)^2/L_y^2]} \quad (5)$$

Δx is the interval between adjacent surface points in the x direction and Δy is the corresponding interval in the y direction. The surface correlation function is defined for a reference point (x_0, y_0) by:

$$C(x, y) = \frac{\langle h(x_0, y_0) h(x_0 + x, y_0 + y) \rangle}{\langle h^2 \rangle} \quad (6)$$

The weights given in Equation (5) yield the following surface correlation function:

$$C(x, y) = e^{-[x^2/L_x^2 + y^2/L_y^2]} \quad (7)$$

This method allows generating different realizations of random surfaces sharing the same statistical parameters (see Figure 7d for an application example).

2.3. Implementation Example for Generation of an Individual Gaussian Rough Surface

The parameters to be chosen as inputs for the parametrization of a Gaussian rough surface are:

- The RMS surface height: $\sigma = \left[\langle h^2 \rangle\right]^{\frac{1}{2}}$. As the height is defined with zero mean ($\langle h \rangle = 0$), the standard deviation $\sigma = \left[\left\langle \left(h - \langle h^2 \rangle\right)^2 \right\rangle\right]^{\frac{1}{2}}$ of the height is equal to the RMS height $rms(h) = \left[\langle h^2 \rangle\right]^{\frac{1}{2}}$: $\sigma = \left[\left\langle (h - \langle h \rangle)^2 \right\rangle\right]^{\frac{1}{2}} = \left[\langle h^2 \rangle\right]^{\frac{1}{2}} = rms(h)$.
- The correlation lengths, L_x and L_y, along the local axes x and y.

The number of mesh points of the rough surface along the local axes x and y has then to be automatically fixed by the chosen mesh algorithm. The previously described Gaussian rough surface model is currently implemented for planar and cylindrical surfaces in 3D computations, and for their linear cross-sections in 2D, in the CIVA software [28] (platform for non-destructive evaluation simulation [29–33]). Examples of generated surfaces for small specimens with relatively high roughness are shown in Figure 1.

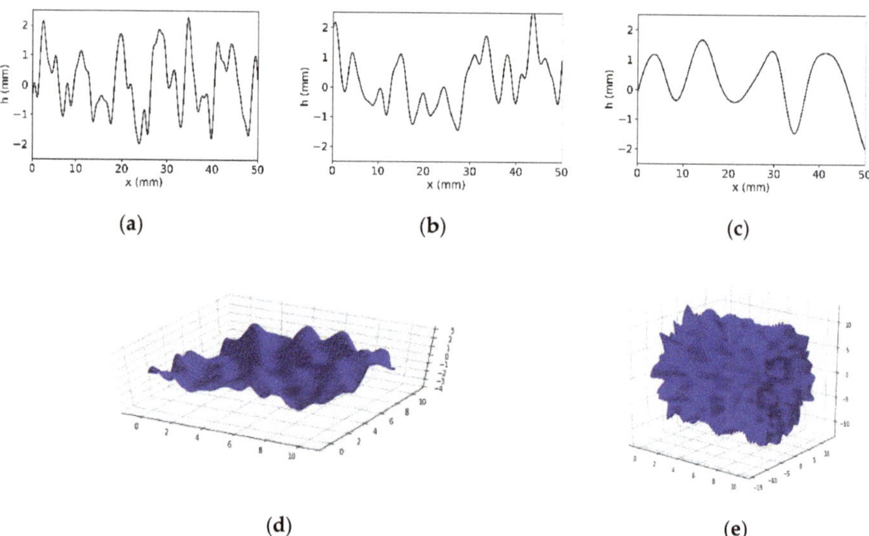

Figure 1. Exaggerated examples of rough surfaces generated for CIVA calculations with standard deviation of the height set to 1 mm for different kinds of rough surfaces: (**a–c**) linear profile for respective ratio $L/\sigma = 1$, $L/\sigma = 1.5$ (case of Figure 6) and $L/\sigma = 4.3$ (case of Figure 11); (**d**) planar for $L/\sigma = 1$; (**e**) cylindrical for $L/\sigma = 1$. Scales in mm.

3. Historical Approximations

This section deals with the perturbation and Kirchhoff approximations. Both have been extensively applied to rough surfaces and have their own strengths and weaknesses. Other approaches will be dealt with in later sections.

Figure 2 shows the geometry used in this section for rough surface scattering.

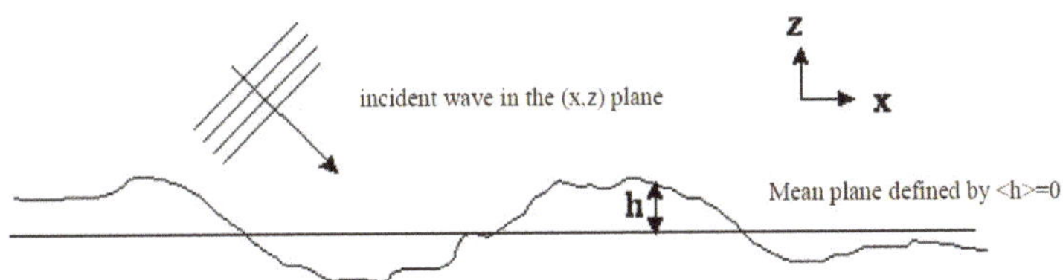

Figure 2. Variations of heights around an average plane.

We consider a rough surface whose roughness is defined by z = h(x,y) where h is the height variation above or below the mean plane, z = 0. The xz plane contains the incident wavevector.

3.1. The Perturbation Theory

The perturbation theory (PT) assumes some restrictive hypotheses on the local height h and gradient ∇h at points of a rough surface:

$$k|h| << 1 \tag{8}$$

$$|\nabla h| << 1 \tag{9}$$

k being the incident wavevector modulus.

It is assumed that if previous conditions are satisfied, the roughness is weak compared to acoustic wavelength and slowly varies. It expresses the scattered field at an observation point r as an expansion series:

$$\psi(r) = \psi^{inc}(r) + \sum_{n=0}^{\infty} \psi_n^{sc}(r) \tag{10}$$

where ψ^{inc} is the incident wave field and ψ_n^{sc} is the nth-order approximation to the scattered field. The $n = 0$ term is the scattered field from a smooth surface. Then the main principle of the perturbation Theory consists in assuming that the boundary conditions may be expanded in a Taylor series about the mean plane z = 0. Mathematically, the boundary conditions on the rough surface expressed as:

$$f(x, y, z) = 0 \tag{11}$$

where f can be, for instance, the acoustical pressure for Dirichlet conditions—which can be expanded in a Taylor series about the mean plane z = 0 as follows:

$$f(x,y,z)|_{z=h(x,y)} = f(x,y,z)|_{z=0} + h\frac{\partial f}{\partial z}|_{z=0} + \frac{h^2}{2}\frac{\partial^2 f}{\partial z^2}|_{z=0} + \ldots. \tag{12}$$

Some applications of the perturbation theory consider first scattering from canonical rough surfaces [18]. For instance for the Lamb's problem (scattering from a surface excited by a load force applied at one particular surface point) [34], the solution is expressed as convolution integrals evaluated only for a triangular irregularity of small height compared to its width and to the wavelength. The application of PT to random rough surfaces leads to consider separately a mean coherent contribution and an incoherent one. For the coherent part, reflection and transmission coefficients are determined. For the incoherent part, the mean intensity of the scattered field is calculated [35]. The determined first-order PT term can be substituted back into the Helmholtz scattering formula to iteratively deduce higher order terms.

Blakemore [36] has then presented a Perturbation Theory at the first order for a fluid/solid interface. PT simulations in backscattering for rigid and elastic solids are very close except at the vicinity of the Rayleigh angle as shown in Figure 3.

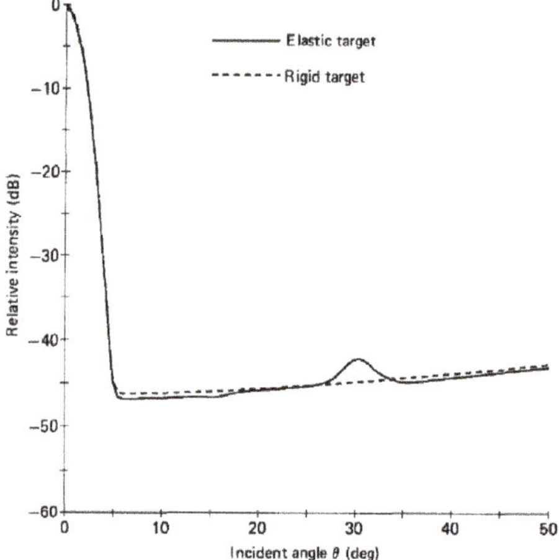

Figure 3. Blakemore's Figure 8 [36]. Backscattered intensity with the perturbation theory on a steel plate immersed in water. The used parameters are respectively frequency = 2.25 MHz, angular beam width = 3°, observation distance = 300 mm, RMS surface height σ = 10 µm and the two correlation lengths of the rough surface both equal to 100 µm. Reproduced with permission from [36], Elsevier, 1993.

In this region a small peak in the intensity of the incoherently scattered field is predicted, consistent with experiment. Nevertheless, in the case of the proposed experimental validation in pulse echo configuration [37] (Figure 4), Kirchhoff prediction (not shown here) are about 3 dB lower than PT one at backscattering angles theta = 40°, and 5.5 dB lower at theta = 70°.

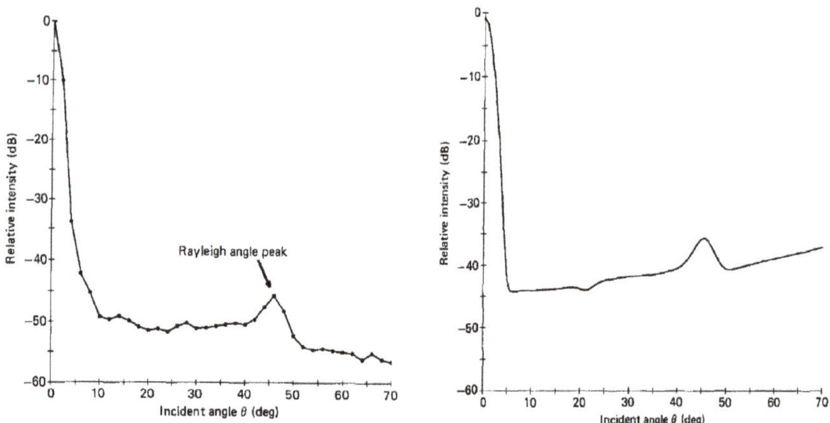

Figure 4. Blakemore's Figures 2 and 11 [36]. Experimental validation of Blakemore's perturbation theory at first order: at left measurements from de Billy et al. [37] on a brass plate immersed in water; at right, Blakemore's results with frequency = 2.2 MHz, angular beam width = 3°, observation distance = 300 mm, RMS surface height σ = 14.9 µm and the correlation lengths L_x = 90 µm and L_y = 105 µm. Reproduced with permission from [36], Elsevier, 1993.

Some applications and references of PT for scattering from fluid/fluid, fluid/solid interfaces and fluid/stratified layers [38–40] for inhomogeneous media [41,42] can be found in the literature, generally for ocean applications. An acoustic model in the time domain based on first-order perturbation theory for scattering from layered sediments has recently been proposed [43].

Only a few authors have been interested in the scattering from canonical non-planar rough targets. Notably Bjorno [44,45] recently investigated the scattering problem for an elastic rough sphere. He developed in that case a second-order perturbation model. He compared the scattering patterns obtained versus the adimensional wave-number ka and for different values of the ratio of the rms height to the sphere radius with its perturbation solution to the classical acoustic scattering solution for a smooth sphere [46,47]. Measurements of the backscattering response of a cast iron sphere versus ka led to a good experimental validation of the proposed model: the roughness yields a decrease in the backscattered amplitude notably for large ka values. The accuracy is limited to small values of the ratio rms height to wavelength.

Note also, for instance, a recent development in electromagnetism of the perturbation theory for a slightly deformed cylinder [48].

We also have to cite the phase perturbation method [49,50] in which the phase (rather than the field) is expanded in a series of different orders in kh. This theory exhibits links with both PT and Kirchhoff [51]. A validation [52] of the phase perturbation method has been proposed and has consisted in comparing its results to numerical ones in a region where PT and Kirchhoff are found ineffective. The phase perturbation method is found to work well in such a region except for low grazing observation angles. However, applications of this new theory seem not numerous and are more focused on radar cross sections calculations [53].

Although the perturbation approach can take into account of the scattering and re-radiation of Rayleigh waves, it appears possibly more restrictive than the Kirchhoff method in its range of applicability. The perturbation method is notably an approach valid for low roughness (when compared to acoustic wavelength). Several validations of the PT are provided in the next sections; they also show that PT is rather accurate far from the specular direction.

3.2. The Kirchhoff Approximation

3.2.1. Theory and History

The main hypothesis of the Kirchhoff approximation (also referred to as the physical optics approximation) is to assume that each point of the surface behaves as if it belonged to the infinite tangent plane to this surface. Then it is practical to employ geometrical acoustics to approximate the exact fields on the local surface: the field on the illuminated part of surface is the sum of the incident field and the field reflected by the local tangential plane. This supposed surface field is then propagated far from the surface using a propagator (called Green function). Indeed the scattered far field is expressed by the way of an integral over the flaw surface (the Kirchhoff–Rayleigh integral) whose integrand is—for the example of Dirichlet conditions—the product of the Green function by the derivative of the geometrical field on the flaw surface. For complex incident fields or scatterer shapes, this integral is evaluated numerically.

The Kirchhoff approximation has been extensively used to model scattering of waves in optics (early introduced by Meecham [54]), electromagnetism [55,56], acoustics, etc. In both acoustics [57] and elastodynamics [58,59], the Kirchhoff approximation also has the advantage in dealing with anisotropic materials and impedant surfaces [57].

Note that when doing an asymptotic evaluation of the KA integral on a finite surface, the KA resulting leading terms describe the reflected waves, and the other terms describe diffracted waves by the surface edges which have the same nature (cylindrical or conical wave fronts) but not the same amplitude as those predicted by the geometrical theory of diffraction (GTD) [55,60]. The main advantage of the Kirchhoff approximation is to provide finite results for the scattered field in the whole space, contrary to GTD [61–65]. One of the main inconveniences of the Kirchhoff approximation (KA) is its incorrect quantitative prediction of edges diffraction, since the approximation of the surface field by the geometrical field is invalid near the edges. To handle this drawback, the physical theory of diffraction

(PTD) has been proposed in electromagnetism [66], acoustics [59,67] and elastodynamics [68–70] and compared to other uniform scattering models [67,70,71].

In the literature, the Kirchhoff approximation has been studied by different authors. A review has been carried out by Ogilvy [18]. The first one to introduce the Kirchhoff approximation is Eckart [72].

Ogilvy [27] has well described the theory for 1D rough profile (Figure 2) and Dirichlet conditions (soft interface) extending Eckart's works. De Billy et al. proposed a Kirchhoff-based theory extended to shadowing effects for the backscattering of acoustic waves by randomly rough surfaces [37]. De Billy et al. employed in fact the potential-method formulation developed by Welton which showed by iterating the scattering integral equation that the scattered pressure can be expressed as a series: the first term corresponds to the singly and the others to the multiply scattered field components. De Billy et al. considered only the first term. Their experimental validations of their model using Gaussian rough (liquid–solid interfaces) surfaces have shown that the experimental measurements of the intensity backscattered by randomly rough interfaces (liquid–solid) are in good agreement with the theory until 40° backscattering angles (with respect to the normal to the mean plane). Nevertheless, an anomaly has been observed in Kirchhoff's predictions for the Rayleigh's angle of incidence. In fact, to model the impedance surfaces used in their experimental validations, they simply add a reflection coefficient as a factor inside the Kirchhoff integral, and for backscattering the local reflection coefficient is the one at normal incidence for a plane surface: this formalism does not take into account any mode conversion in the elastic medium and is unable to predict any perturbation at the different critical angles. Following de Billy's works which studied the Kirchhoff approximation using monochromatic plane waves, Imbert [73] has proposed a Kirchhoff model in time domain for 2D rough profiles and Neumann conditions (rigid interface).

We can sum up the main steps of Kirchhoff approximation application as follows:

1. Use of the Helmholtz equation: the scattered field is:

$$\psi(r) = \psi^{inc}(r) + \int_{S_0} \left(\psi(r_0) \frac{\partial G(r, r_0)}{\partial n_0} - G(r, r_0) \frac{\partial \psi(r_0)}{\partial n_0} \right) dS_0 \quad . \tag{13}$$

The equation is also called Green theorem and $G(r, r_0) = \frac{e^{ik|r-r_0|}}{4\pi|r-r_0|}$ is called the Green function, where r_0 is a current point on the scattering rough surface S_0 and r the observation point. The field in the fluid medium can be obtained thanks to the knowledge of its value on the surface S_0.

2. Boundary Conditions on the rough surface: Ogilvy [27] considered for instance Dirichlet boundary conditions (soft case) so that $\psi(r_0) = 0$. So one term in the previous Green integral disappears:

$$\psi(r) = \psi^{inc}(r) + \int_{S_0} \left(-G(r, r_0) \frac{\partial \psi(r_0)}{\partial n_0} \right) dS_0 \tag{14}$$

3. Kirchhoff approximation: the tangential plane approximation enables to approximate the value of $\frac{\partial \psi(r_0)}{\partial n_0}$ inside the previous integral.
4. Far field approximation: the integrand value can be simplified by assuming $kr >> 1$, k being the wave number.
5. Projection of integration on the rough surface on the smooth surface: this projection involves the gradients of the roughness.
6. Incorporation of a shadow function to approach auto-shadowing: Wagner [74] has proposed to incorporate a shadow function inside the Kirchhoff integral to take into account shadowing effects.

Note that the Kirchhoff approximation is mainly used to predict far-field scattering; a model has been proposed to evaluate the scattering pattern in near-field [75] after the interaction of an acoustic wave with a rough interface.

3.2.2. Validity

Historically KA has been considered—in the point of view of its local tangential plane assumption—as accurate when the rough surface can be seen planar by the incident wave. In Ogilvy's review [7], KA is thus said valid for $kr_c \cos^3 \theta i >> 1$, with r_c the surface local radius of curvature and θi the incident angle. This historical criterion forgets that multiple scattering and shadowing effects are not taken into account in KA leading to large errors for grazing incidences or higher roughness.

KA prediction has been compared to Integral Equation (IE) results by Thorsos [21] for 1D rough profiles. Thorsos proposed to use the surface correlation length L as main parameter for defining the Kirchhoff validity away from the low grazing angle region: KA is said to be accurate for a correlation length greater than the wavelength or even for lower correlation lengths if shadowing effects are not important. We notice this first criterion proposed by Thorsos at this step is too simple and not involve the rms height. In the low grazing angle region some examples of validation with comparison to another methods are given in next Section 4; at low grazing angles, additional validity dependence enters through the RMS slope which is for Gaussian rough surfaces: $s = \tan\gamma = \sqrt{2}\frac{\sigma}{L}$ i.e., the rms height is now involved. However, at shorter correlation lengths, the needed shadowing correction proposed by Thorsos becomes greater to improve Kirchhoff prediction at low grazing angles in the case of backscattering.

Note that in order to improve its prediction for low grazing angles, Chapman [76] provided an improved Kirchhoff formula (using a smoothing special function) notably by including the correlation length of the scattering surface into the reflection coefficient. The modification impact is not noticeable since shadowing and multiple scattering will dominate for low grazing angles. Chapman showed that the PT and KA results differ except in the limit of infinite correlation length (smooth surface).

Shi et al. [77] recently proposed for elastic materials (also for shear waves [78,79]) and for compressional waves a different surface parameter than the surface correlation length L used by Thorsos, a function σ^2/L of both L and σ, to define a criterion when the KA use is valid. They found that, with a normal incidence angle $\theta i = 0°$, KA is valid with a scattering angle $-70° < \theta s < 70°$ and when $\sigma^2/L \leq c$, the constant upper bound c being in two dimensions 0.20 λp and in three dimensions 0.14 λp (λp being the compressional wavelength). Figure 5 shows a comparison between KA and a finite element (FE) method for a 2D case when $\sigma^2/L \sim c$. A small incidence angle of 30° can improve the KA accuracy when σ^2/L exceeds c as in the case shown in Figure 6. We can however regret that the numerical evaluation of this proposed criterion was limited to small incidences angles: however, this criterion is likely to be suitable for many NDE configurations for which incidence is close to normal. A study of this criterion near grazing incidence should also be useful.

As said in the introduction, another different KA validity criterion has been proposed for radar applications [22] and is also based on numerical calculations: it is given by Equation (1) in [22] (see also their Figure 3, Kirchhoff being called the scalar approximation) and ensures a KA validity for L^2/σ more than a constant times the wavelength and a condition $kL > 6$ is also associated. We find the condition on L^2/σ very disbelieving and in contradiction with the previously cited studies and also that given by Voronovich ($\sigma/L < C$ and large kL; see Figure 1.1 in [20]). The condition $kL > 6$ is also found in other references as [52]. Notably, Figure 2 in [52] represents the stated validity of first-order perturbation theory and KA in $k\sigma$–kL space obtained for bistatic scattering cross sections simulation: PT is said valid for small $k\sigma < 2$ to 4 depending on kL value, KA valid for $kL > 6$ and both KA and first-order PT needs $\sigma/L < C$ (C being a constant value).

In our opinion, this KA condition $kL > 6$ is too restrictive for many applications, such as non-destructive evaluation, which often employs near specular observation inspection. First, we previously study or carry out numerical comparisons of KA simulation for cracks of finite size L inspected with longitudinal waves in solids with numerical methods: the method of optimal truncation (MOOT) [80] and a finite element method (FEM) [81,82]. These comparisons with MOOT [83] and FEM [69,84] both show that KA is valid for near specular observation inspection for $kL > 2$. Shi et al. [77] did not propose a validity criterion on kL (but only the condition on σ^2/L); they nevertheless obtained an acceptable disagreement of 2dB (see their Figure 10 [77]) for $kL = \pi/2$ (corresponding to L = λp/4) and for an important roughness σ = λp/4 = L equal to the correlation length. It is a pity that that they did not provide numerical validations below $kL = \pi/2$.

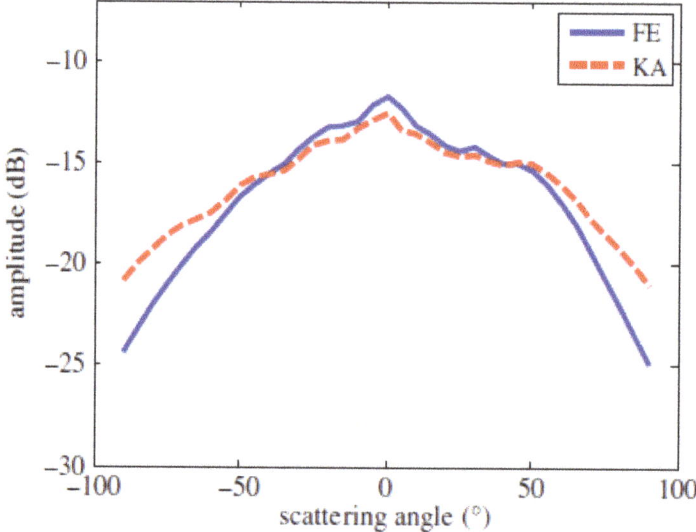

Figure 5. Shi's Figure 11 [77]. Scattered amplitude versus observation angle using 2D finite element (FE) and Kirchhoff approximation (KA) simulations, for θi = 0°, $\sigma = \lambda$p/3, $L = 0.6\ \lambda$p. Reproduced with permission from [77], Royal Society, 2015.

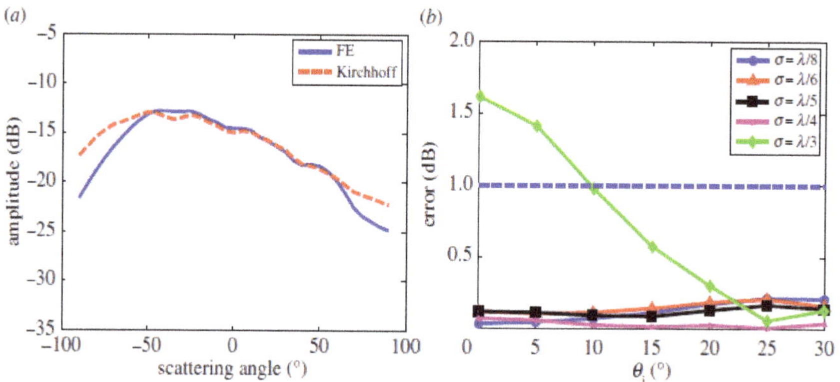

Figure 6. Shi's Figure 12 [77]. (**a**) Shi's Figure 11 [77]. Scattered amplitude versus observation angle using 2D FE and KA simulations, for θi = 30°, $\sigma = \lambda$p/3, $L = \lambda$p/2 < 0.6 λp. (**b**) Error between 2D FE and KA with respect to θi in the specular direction for different values of σ. Reproduced with permission from [77], Royal Society, 2015.

Our conclusive opinion is that for many applications using rather near specular observation and non-grazing incident angles, two criteria can be used for KA validity: $\sigma^2/L \leq C\lambda$p and kL > 1 or 2. For more important observation angles to the normal the criterion on kL and the constant C value are likely to be a little more restrictive: Zhang [85] using 2D FEM calculations recently showed that KA can be considered accurate for kL > π and for incidence and scattering angles over the range from −80° to 80°.

3.2.3. Elastic Targets and Applications

The Kirchhoff approximation models bulk waves, and does not take into account surface acoustics waves (SAW) as Rayleigh waves neither head waves. Head waves led to recent researches [70,86–88] and a modification [89] of the Kirchhoff approximation is henceforth proposed for head waves account. Head waves on irregular surfaces can also be modelled [90].

The Kirchhoff approximation for elastic rough targets is proposed by Ogilvy [13] and then revisited by Bjorno [91] and Dacol [92]. The Kirchhoff approximation validity have also been specifically considered in the particular cases of rigid and soft targets in a modeling study [93] of the reflected field including multiple scattering. One example of KA application to impedance surfaces [57] for non-destructive evaluation (NDE) is the modeling of ultrasonic telemetry of immersed targets [59,94].

We report here an application of KA simulation to impedance rough surfaces for NDE. Ultrasonic inspection of a planar steel component immersed in water (water path 25 mm) at frequency = 2 MHz (wavelength λ = 0.75 mm) is simulated with the Kirchhoff approximation for elastic rough entry surfaces of varying roughness using CIVA software inspection simulations tools [57,95,96]. A phased-array (represented in yellow in Figure 7a,b) is used to emit an incident beam directed at 117° with respect to the surface thanks an emission delay law (applied in red in Figure 7a to only some left elements) and all elements receives the scattered waves without delay law in the angular domain (118°, 58°).

In Figure 7a,b, the true B-scans (true-to-geometry imaging) are surperimposed to the specimen in color codes; they represent the echographic signal versus time for all elements in reception (corresponding to scattering angles in the angular domain (118°, 58°)). Figure 7c shows that increasing the roughness deteriorates the specular reflection echo and causes appearance of echoes at non-specular angles. Figure 7d shows that the scattering directivity patterns obtained for three different generated rough surfaces of the same statistical properties are slightly different.

A recent study [97] has compared the accuracy of KA and PT using a finite element (FE) model for fluid and elastic ocean bottoms. For a fluid bottom, KA is close to FE with maximal gap of 2 dB except for small grazing incident angles for which PT is more effective. PT behaves globally well except near specular observation direction notably for small grazing incident angles. For the elastic case, the disagreement between KA and FE increases to 5 dB and KA is inaccurate for incident angles less than the longitudinal critical angle: the interferences between transversal, longitudinal and surface waves are not taken into account in KA. The PT prediction is generally good, but another limitation appears in the elastic case for incident angles near the longitudinal critical one. However, we regret the provided validations are established only for a small roughness k σ= 0.6 favorable to PT.

3.2.4. Periodic Surfaces

As it will be discussed in the Section 5, Perspectives, it can also be useful to model scattering from periodic surfaces—such surfaces can notably appear in additive manufacturing parts or in oceans. For the case of periodic surfaces, one exact solution is the classical Rayleigh–Fourier method (RFM) [98]. Impinged by a plane wave, a periodic surface leads to Bragg scattering with a finite number of plane wave propagation angles in far field. A very recent study evaluates the validity of the Kirchhoff approximation (KA) for the scattering problem from a sinusoidal surface—mimicking the ocean one—with comparison to the RFM solution [99]. An asymptotic evaluation of the Kirchhoff integral which is consequently equivalent to geometrical acoustics [100] provides a ray theory which is also evaluated to better understand the Kirchhoff prediction. This notably shows that Kirchhoff has the advantage to exhibit a smooth behaviour near caustics generated by the sinusoidal surface contrary to geometrical acoustics. The KA validity is also studied versus the observation range (projected distance between the source and receiver points): for a short range, KA is very close to RFM. At moderate range, KA sometimes over-estimated the peak pressure due to interferences between several rays. For longer ranges, the KA disagreement increases strongly. This study [99] is very useful for better understanding the KA limitations due to multiple paths arrivals and grazing incidence/observation. Nevertheless, KA remains a very useful tool even for periodic surfaces when source and observations are not too far.

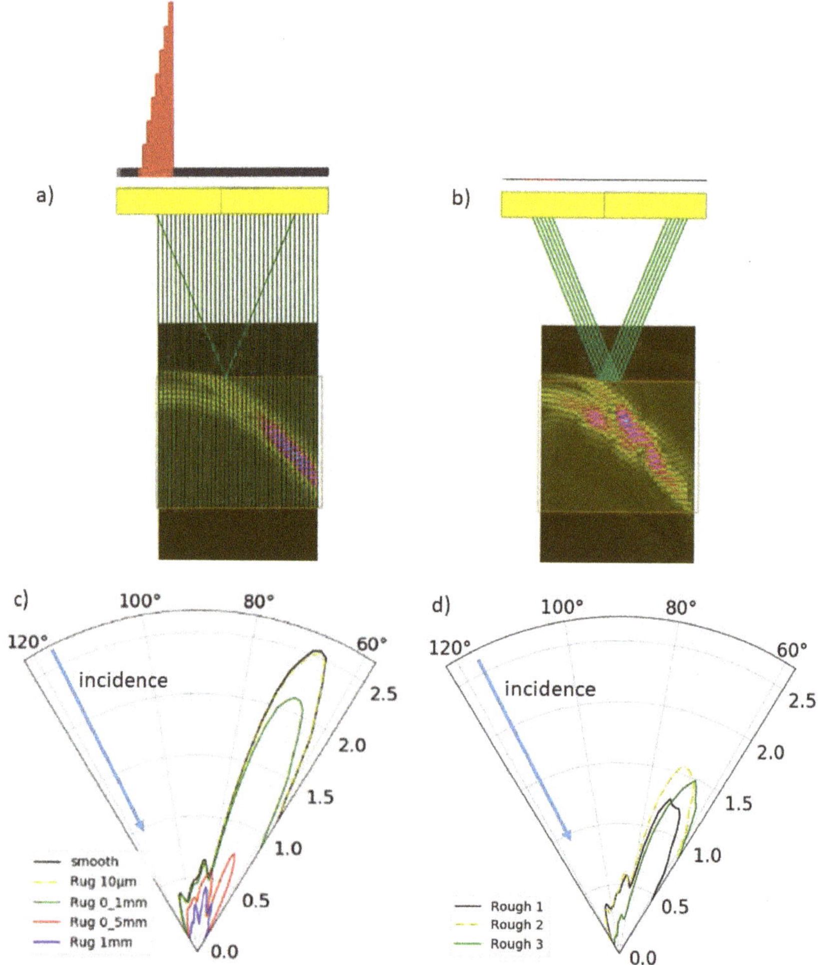

Figure 7. Ultrasonic inspection of a planar steel component (orange box) immersed in water (water path 25 mm) at frequency = 2 MHz (wavelength $\lambda = 0.75$ mm): Bscans simulated with KA for a (**a**) smooth and (**b**) a rough entry surface (RMS surface height $\sigma = 1$ mm and correlation lengths $L_x = L_y = 10$ mm). Scattering patterns (maximum of the echographic signal versus observation angle) (**c**) for different roughnesses (smooth, 10 μm, 0.1, 0.5 and 1 mm) and (**d**) for three different generated rough surfaces of the same statistical properties (roughness $\sigma = 0.2$ mm and correlation lengths $L_x = L_y = 10$ mm).

4. More Modern Approaches with Larger Domains of Validity

We have presented the two classical approximations: the perturbation theory (PT) and the Kirchhoff approximation (KA). In the view of main application to scattering from ocean, it is important to cite the two-scale roughness theories [101–104]. They consist in assuming that scattering can be considered as the combination of two phenomena: that produced by large facets and well approximated by KA and that due to small roughness and approximated by PT. One major inconvenient of this technique is that it requires to fix a wavenumber parameter [105] to separate the two scattering domains.

The small slope approximation, described in the next section, is a much more convenient technique to link the classical approximations and it reduces to their results in some limiting cases.

In this section, some sound propagation models which have been notably devised for underwater acoustic applications and providing a better validity notably for grazing observation angles are studied. Underwater acoustic propagation depends on many factors. When sound propagates underwater the sound intensity decreases when increasing the propagation distance. Propagation loss is a quantitative measure of the reduction in sound intensity between two points, normally the sound source and a distant receiver. Modeling the acoustic reflection loss [106] at the rough ocean surface needs simulating forward scattering for incident grazing angles and is of interest for sonar especially for shallow oceans or oceans with a surface duct (sound waves trapped in a duct layer below the sea surface) [106]. Several models described hereafter (small slope approximation—SSA, and parabolic equation—PE) were notably designed to handle grazing angles. The integral equation (IE) method is also used as reference solution for validating SSA and comparing it to other approximations as KA and geometrical optics (GO).

4.1. Small Slope Approximation (SSA)

The SSA technique was described first by Voronovich [107–109]. An alternative small slope approximation has originally been developed by Urusovskii and then studied by McDaniel (1994) [110]. All these methods attempt to exploit the assumed smallness of the surface slope without restricting the RMS surface height to wavelength ratio.

The SSA model from Voronovich has been investigated by Thorsos and Broschat [111,112]. SSA consists in introducing a roughness correction to the Helmholtz integral. This correction is expanded into a functional Taylor series in slope whose terms depend on the integral of the wave number spectrum of the surface height function. The approximation of each order of SSA is based on the approximations of previous orders. Owing to its principles SSA takes into account shadowing and diffraction.

Thorsos and Broschat showed that, at lowest order the small slope approximation of Urusovskii/McDaniel is limited to forward scattering [110] but that the lowest order term in Voronovich's SSA was incorrect in McDaniel's SSA.

SSA tends to the perturbation approximation for small k σ (adimensional wave numbers in terms of RMS surface height σ). The second order SSA (SSA (2)) reduces to the Kirchhoff approximation (KA) at high frequencies, hence does not take into consideration the shadowing effects. The predictions of SSA(2), SSA(3), and SSA(4) are compared [112] with Monte Carlo integral equation (IE) calculations devised earlier by Thorsos [21], and with other approximations: the perturbation approximation, the Kirchhoff approximation (KA), and geometrical optics (GO). The scattering geometry is described in Figure 8.

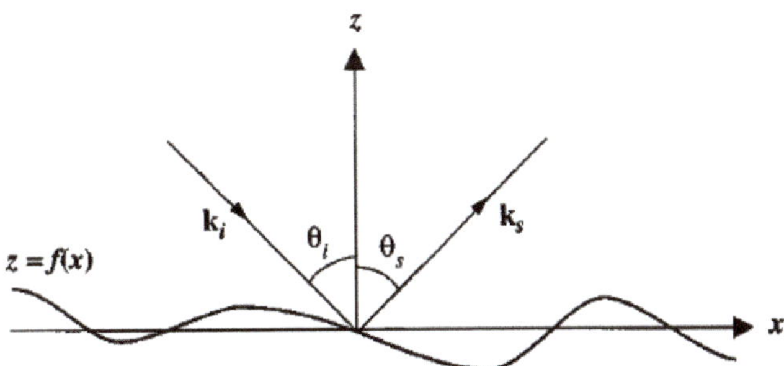

Figure 8. Figure 1. Thorsos' Figure 1 [112]. Scattering geometry. Reproduced with permission from [112], AIP Publishing, 1997.

SSA and IE scattering predictions are presented in Figure 9 for respective k σ and kL 0.5 and 4.0, for 45° incident angle and RMS slope s = 0.177 ($s = \sqrt{2}\frac{\sigma}{L}$ for Gaussian surface), and the RMS slope angle $\gamma = \tan^{-1} s$ is 10°. The contribution of specular reflection—leading to a peak in the IE prediction—has not been integrated in the SSA models.

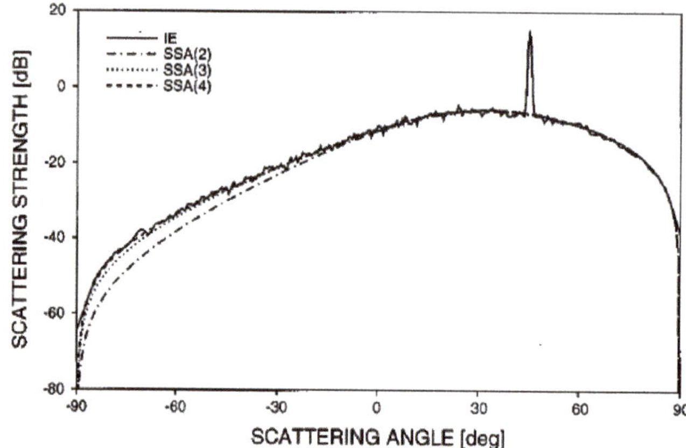

Figure 9. Thorsos' Figure 2 [112]. Comparison of the second-order SSA (2), third-order SSA(3), and fourth-order SSA(4) small slope approximation and Monte Carlo integral (IE) scattering strengths (=10 log of the scattering cross section) for a modest rms surface slope angle. Here, k σ = 0.5, kL = 4.0, the RMS slope angle γ is 10°, and the incident angle θ_i is 45°. Reproduced with permission from [112], AIP Publishing, 1997.

In Figure 10, k σ and kL have both been increased by a factor of three to 1.5 and 12.0, respectively. The increase in the surface roughness (resp. correlation length) causes disappearance of the specular peak (resp. decrease of the backscattered strength). A higher SSA order improves the prediction in backscattering.

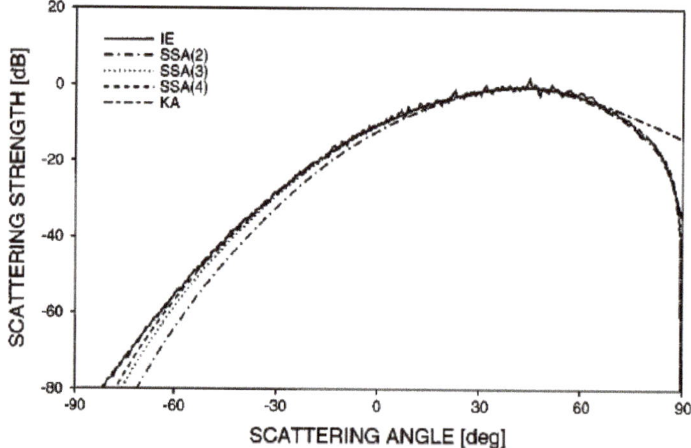

Figure 10. Thorsos' Figure 3 [112]. Comparison of SSA(2), SSA(3), SSA(4), Kirchhoff approximation (KA) and IE scattering strengths. Here k σ = 1.5, kL = 12.0, γ is 10° and θ_i is 45°. Reproduced with permission from [112], AIP Publishing, 1997.

According to Thorsos [21], the Kirchhoff approximation KA is accurate away from low grazing angles for rough surfaces smooth on the scale of a wavelength, assuming small surface slopes. For a Gaussian roughness spectrum, the KA is said to be generally accurate for kL >6, that is when $\frac{L}{\lambda} > 1$. We have seen in Section 3.2.2 the recent definition of a more precise criterion for the KA validity. In the following of this section the SSA and KA are thus considered for several examples with kL > 6.

In Thorsos' previous Figure 3 [112], KA is valid away from grazing angles, whereas PT (not shown) is not. The SSA(2) result reduces to the KA result in the specular region; the third and fourth-order results do as well. KA behaves better than SSA for backscattering, but it is the contrary in the forward direction at grazing angles.

Finally, several cases have been examined beyond the region of validity of the fourth-order Perturbation Theory PT(4) and of KA. For example, in their Figure 6 [112] since k σ = 1, PT(4) breaks down whereas KA is globally correct and SSA(4) the more realistic (Figure 11).

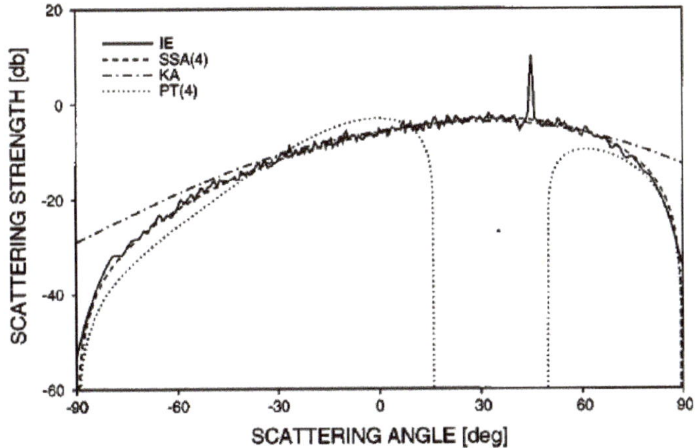

Figure 11. Thorsos' Figure 6 [112]. Comparison of the SSA(4), KA, PT(4) and IE scattering strengths for a case when neither PT(4) nor the KA is accurate. Here, k σ = 1.0, kL = 4.3, γ = 18.2°, and θ_i = 45°. Reproduced with permission from [112], AIP Publishing, 1997.

Other validations showed that the fourth-order SSA (SSA(4)) mimics some shadowing phenomena and the SSA results are globally accurate for all observation angles for incident angles up to 80°, but the validity decreases with the incident angle.

Another study [113] of the Bragg scattering from surface waves by a rough interface recently provided comparisons between the Kirchhoff approximation, the perturbation theory and the integral equation method; Kirchhoff leads to results relatively close to the integral equation method in the studied configurations (in forward and backscattering).

Voronovich (see Figure 1.1 in [20]) argued that SSA is valid for σ/L less than a constant value. According to him, KA and PT also require this condition but PT needs in addition a small kσ and KA a great kL.

SSA has also been developed for rough solid elastic interfaces [114]. Berman [115] showed that it provides an effective solution compared to Monte-Carlo integral equations simulations for grazing incident angles, whereas Kirchhoff leads to an overall important underestimation. SSA and first-order PT give rather good predictions and exhibit minima or peaks at the Rayleigh angle that can be exaggerated or discontinuities at critical angles which are not physical. The authors noted that Kirchhoff is rather smooth or does not have the valid behaviour for these particular angles: it is logical because the associated phenomena (Rayleigh or head waves) are not accounted on the simplified Kirchhoff theory (see also the perspectives section). We nevertheless notice that Kirchhoff yields an overall rather good prediction versus scattering angle despite some quantitative discrepancies. We also emphasise that one comparison shows an overestimation by first-order PT of the specular amplitude, which is a strong inconvenient for numerous applications as NDE.

SSA for layered fluid seafloors has only been studied recently. A study published this year [116] reviewed three different approximations. Jackson [117] has proposed an acoustic SSA method for rough water–sediment interfaces with a fluid layer of finite thickness overlying a semi-infinite fluid layer. A second method consists in using in acoustics the SSA developed for the general layered case in electromagnetics [118–120]. A third approach is developed in [116] by translating usual small-slope ansatz from k-space to coordinate space. It is then shown in [116] using different examples that the different approximations give different results in the case of strong internal reflections and also differ from Kirchhoff in specular direction. In an example with a mud layer overlying a harder basement, the first approximation differs strongly from the two others and consequently seems erroneous. Unfortunately, no numerical validation is provided to define the best ansatz.

4.2. Parabolic Equation (PE)

Acoustic losses at the rough ocean surface can be simulated utilizing a propagation code which can use as input an explicitly defined roughness profile. The PE (Parabolic equation [121–123]) model Ramsurf [124] is often used and enables to model shadowing from the surface roughness. Ramsurf is based on the split-step Padé method [121] (a pseudo-spectral numerical method for solving nonlinear partial differential equations).

Firstly, based on the rough ocean surface of Gaussian spectrum, the validity of Ramsurf acoustic propagation model has been studied [125] by comparison with measurements after an explosion at receivers located on the ocean bottom around 1000 m depth. The height for the sea surface is from 1.5 m to 2 m. Transmission loss curve under a smooth sea surface and that under the rough sea surface are simulated respectively by Ramsurf. The sea surface reflection loss caused by the surface roughness is about 4 to 6 dB. The transmission loss simulated by Ramsurf model is very close to the experimental one. Kirchhoff approximation (KA) and small slope approximation (SSA) abilities to simulate the reflection loss are also compared: SSA and Ramsurf match well for very small grazing angles (<6°) for which KA breaks down. We nevertheless notice that the gap between KA and Ramsurf becomes less than 3 dB above 6°: KA is thus inaccurate only in a very small region.

Another study [126] focuses on the modeling of underwater acoustical reflection from a wind-roughened ocean surface and compares the surface loss between KA, a small-slope model SSA(2) and a stochastic modeling using Ramsurf. SSA(2) matches well with Ramsurf for the entire range of frequency and wind speed combinations studied (1.5–9 kHz, 5–12.5 m/s) and for sound incidence grazing angles from 1 to 11°. KA can be used effectively from an angle about 2.5 to 7°.

An extension of the parabolic equation method to deal with forward scattering has recently been proposed for both Dirichlet and Neumann boundary conditions [127] but future works are needed to make the connection between forward and backward 3D scattering and treat other boundary conditions.

5. Summary, Discussion and Perspectives

A review of acoustic wave scattering from rough surfaces has been done. Several analytical models proposed in the literature have been described in more detail as well as their validity. We are also focused in specific acoustic applications as scattering from ocean, non-destruction evaluation, medical sciences.

Two classical approximations have been early devised: the perturbation theory and the Kirchhoff approximation.

The perturbation theory (PT) consists in expressing the scattered field in an expansion series using as small parameter the roughness height. This method leads to a major inconvenient: it is valid for low roughness when compared to acoustic wavelength (at first order for small $k\sigma < 2$ to 4 depending on kL value, σ, L and k being respectively the rms roughness height, the correlation height and the wave number). Existing validations also showed that PT is rather accurate far from the specular direction. Consequently, it appears possibly more restrictive than the Kirchhoff method for numerous applications notably non-destruction evaluation (NDE) which usually employs inspection in specular direction.

The Kirchhoff approximation (KA) leads to a simple analytical expression; it is a relatively powerful and inexpensive approximation in terms of calculations and easily extended to impedance rough surfaces. One important fact is that KA is the most used method in NDE for rough surfaces: its implementation of the rough structure echo consists in simply modifying that of the smooth surface; the "rough" KA algorithm can thus simply mesh the smooth mean surface (surface without roughness) in small surface elements and then modify the contribution of each mesh point by including some local geometrical parameters of the rough surface.

KA has the advantage of handling surfaces of roughness smaller than a constant depending on the wavelength and on the correlation length with a relatively satisfying validity except at the vicinity of grazing observation and critical angles. Indeed we have connected and merged recent studies of the KA validity and conclude that for many applications using rather near specular observation and nongrazing incident angles, two criteria can be used for KA validity: $\sigma^2/L \leq C\lambda p$ (C and λp being a constant value and the wavelength) and kL >1 or 2. If we do not want to limit the KA validity to near specular observation, the previous criteria are likely to be a little more restrictive to reach the KA validity for incidence and scattering angles over the range from −80° to 80°(kL > π is for instance needed)

It is well admitted that KA validity for application to rough surfaces decreases for grazing angles and the region of inaccuracy in terms of scattering angles is much more important for a fluid/solid interface. One has to pay attention at the KA validity in the vicinity of particular angles (Rayleigh and critical ones). In NDE inspections, these angles are often avoided by the inspector; nevertheless it may be possible to meet them in some geometrical configurations. Improving KA or analytical approximations to better take into account head waves occurring at critical angles is an important challenge already for planar surfaces [128]. Modeling head waves on irregular surfaces is even more complicated and seems intractable using analytical methods [90,129]. Finite elements methods [130] can be useful to model both the head waves generated at critical angles, the bulk waves radiated near the surface and their possible interferences [90,129].

For planar surfaces, some smoothing modifications of KA can be proposed [69] in the vicinity of critical angles for better validity, but they do not simulate head waves. A modification of KA using an alternative integral formula for the Kirchhoff integral has just been developed [89], and it could be extended to scattering from a rough surface. Nevertheless, this perspective has to be tested and validated, since such a model will generate head waves as if they have been created on successive local non-finite tangential plane.

PT and KA are usually seen as complementary, PT being valid for small roughness scales and Kirchhoff for large scales. Several other approximations have been developed in the view of handling the inconveniences of the classical methods (PT and KA) and enlarge the simulation validity. Some of them (phase perturbation theory, two-scales models) found some applications in electromagnetism. A recent study [131] has just began in order to try to improve the two-scale models. We are not convinced that improving these models is the best perspective because they inherently own a certain complexity and there are probably better candidates (evoked in the following) in simulation methods.

The small slope approximation (SSA) is a model based on an integral field equation; it is more efficient than Kirchhoff for grazing angles but is only valid for small surface slopes. Indeed, the SSA is of great importance in ocean applications for modeling scattering from ocean surfaces on which the incidence is generally near from grazing: in such a configuration the SSA accuracy is well established. One main future challenge is to extend this approximation to layered fluid seafloors. In a study published this year [116], different approximations are proposed but they give different results in the case of strong internal reflections and also differ from Kirchhoff in specular direction. Numerical validations are clearly needed to try to find the best approximation among them; another problem is that only one extended from electromagnetism [118] can deal with multiple layers.

The parabolic equation describes all physical effects including shadowing, but this method needs a complex and heavy numerical technique to solve the parabolic equation. The parabolic equation method has recently been extended to forward scattering, but numerous efforts should be necessary in the future to build a more robust model applicable for both forward and backward 3D scattering. Such a strategy can be seen complex when compared to other methods as fully numerical ones which are generic for all scattering directions.

There is also recent progress in methods based on the integral equation; for instance the new development described in [132] leads to impressive results, notably for grazing angles; but the calculation time is still for the moment a limitation by comparison with simpler methods, if we are rather interested in near normal incidence.

Numerical methods are obviously the main rivals of the existing analytical models. For numerical techniques, obtaining 3D simulations with a reasonable computation time remains a challenge. In a very recent study, the boundary element method has been adapted [133] to reduce the 3D calculation time; the proposed adaptation is validated by using approximated models (KA and PT) showing that the classical approximations are also useful for validation purposes.

In the field of NDE, one major future challenge is to model scattering from roughness in components fabricated by additive manufacturing (AM). The fatigue life of AM specimens is very influenced by their surface topography [134]. AM processes for metallic parts are notably the selective laser melting (SLM), selective electron beam melting (SEBM), laser metal deposition (LMD) and also wire arc additive manufacturing (WAAM [135]). The corresponding roughness value of each process is discussed in [136,137]: it is 4–11 μm Ra for SLM, 15–35 μm Ra for SEBM, 10 μm Ra and 200 μm Ra for LMD, and can be very important for WAAM. Note that models are searched for the prediction of surface roughness (see a list of studies in the introduction of [138]), the optimization of process parameters as well as the roughness minimization (as in [138]) for manufacturing using WAAM and milling.

In AM processes there a lot of phenomena causing roughness [139,140], notably the sticking of nonmelted or partially melted particles on the surfaces and the formation of menisci. The latter phenomenon is more pronounced for the WAAM process (see Figure 12, below).

Figure 12. Surface topography of a specimen manufactured by wire arc additive manufacturing (WAAM). Reproduced from [141] (Open access, no copyright).

The balling effect (Figure 13) is a detrimental phenomenon that may result in breaking up the liquid scan track during SLM and produce particles in spherical shapes.

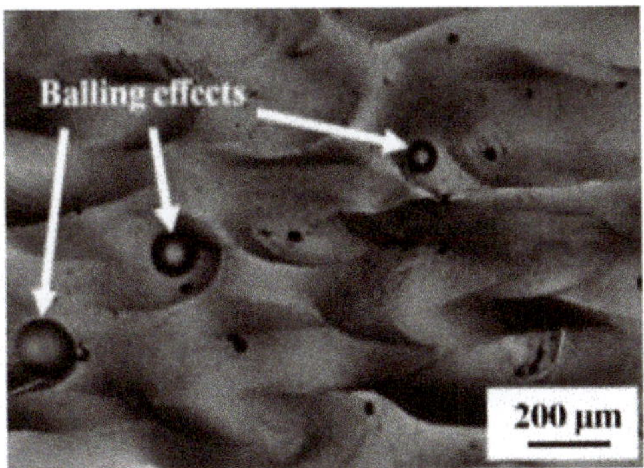

Figure 13. Surface topography of a specimen manufactured by selective laser melting (SLM). Reproduced from [137] (Open access, no copyright).

The rough surface of AM parts can thus exhibit a kind of periodicity and elementary egg-shaped irregularities [142]. Finite element models [143] are likely to be the best solution to both model the bulk macrostructure of such components and their surface roughness—such models have been recently used to model scattering in virtual simulated microstructures of polycrystalline materials either with equiaxed grains [144–146] or transverse isotropic as in austenitic welds [147]. The increasing power brought by the spectral finite elements (SEM) method is noticeable [148–150] in terms of calculation time and could be very useful for roughness with complex shape irregularities; SEM has recently been used for modeling ultrasonic propagation in granular materials, such as concrete [151].

In another domain, a FEM model has just been employed to model the ultrasonic response of bone–dental implant interfaces [5,152] whose irregularities can generate multiple scattering. Such a numerical method also presents the advantage of being easily adaptable in the future, for instance to treat non-linear effects at the contact interface. All the previous examples show the increase of applications for numerical methods.

Author Contributions: Software, V.D. and M.D.; validation, M.D. and V.D.; investigation, M.D.; resources, M.D.; writing—original draft preparation, M.D.; writing—review and editing, M.D. and V.D.; project administration, F.B. All authors have read and agreed to the published version of the manuscript.

Funding: This work was supported through the "Generation IV/Sodium Fast Reactors/ASTRID" program with CEA, EDF, and FRAMATOME funded by the French government.

Conflicts of Interest: The authors declare no conflict of interest. The funders had no role in the design of the study; in the collection, analyses, or interpretation of data; in the writing of the manuscript, or in the decision to publish the results.

References

1. Romanova, A.V.; Horoshenkov, K.V.; Krynkin, A. Dynamically Rough Boundary Scattering Effect on a Propagating Continuous Acoustical Wave in a Circular Pipe with Flow. *Sensors* **2018**, *18*, 1098. [CrossRef] [PubMed]
2. Wang, Z.; Cui, X.; Ma, H.; Kang, Y.; Deng, Z. Effect of Surface Roughness on Ultrasonic Testing of Back-Surface Micro-Cracks. *Appl. Sci.* **2018**, *8*, 1233. [CrossRef]
3. Dingler, J.R.; Boylls, J.C.; Lowe, R.L. A high-frequency sonar for profiling small-scale subaqueous bedforms. *Mar. Geol.* **1977**, *24*, 279–288. [CrossRef]
4. Niu, L.; Qian, M.; Yang, W.; Meng, L.; Xiao, Y.; Wong, K.K.L.; Abbott, D.; Liu, X.; Zheng, H. Surface Roughness Detection of Arteries via Texture Analysis of Ultrasound Images for Early Diagnosis of Atherosclerosis. *PLoS ONE* **2013**, *8*, e76880. [CrossRef] [PubMed]
5. Hériveaux, Y.; Nguyen, V.-H.; Biwa, S.; Haïat, G. Analytical modeling of the interaction of an ultrasonic wave with a rough bone-implant interface. *Ultrasonics* **2020**, *108*. [CrossRef]

6. Ogilvy, J.A. Model for the ultrasonic inspection of rough defects. *Ultrasonics* **1989**, *28*, 69–79. [CrossRef]
7. Ogilvy, J.A. Theoretical comparison of ultrasonic signal amplitudes from smooth and rough defects. *NDT Int.* **1986**, *19*, 371–385. [CrossRef]
8. Ogilvy, J.A.; Culverwell, I.D. Elastic model for simulating ultrasonic inspection of smooth and rough defects. *Ultrasonics* **1991**, *29*, 490–496. [CrossRef]
9. Nagy, P.B.; Adler, L. Surface roughness induced attenuation of reflected and transmitted ultrasonic waves. *J. Acoust. Soc. Am.* **1987**, *82*, 193–197. [CrossRef]
10. Nagy, P.B.; Rose, J.H. Surface roughness and the ultrasonic detection of subsurface scatterers. *J. Appl. Phys.* **1993**, *73*, 566–580. [CrossRef]
11. Nagy, P.B.; Adler, L.; Rose, J.H. Effects of Acoustic Scattering at Rough Surfaces on the Sensitivity of Ultrasonic Inspection. In *Review of Progress in Quantitative Nondestructive Evaluation*; Springer: Boston, MA, USA, 1993; pp. 1775–1782; ISBN 978-1-4613-6233-3.
12. Bilgen, M. Effects of randomly rough surfaces on ultrasonic inspection. *Retrosp. Theses Diss.* **1993**. [CrossRef]
13. Ogilvy, J.A. *Theory of Wave Scattering From Random Rough Surfaces*; Taylor & Francis: Abingdon, UK, 1991; ISBN 978-0-7503-0063-6.
14. Elfouhaily, T.M.; Guérin, C.-A. A critical survey of approximate scattering wave theories from random rough surfaces. *Waves Random Media* **2004**, *14*, R1–R40. [CrossRef]
15. Ishimaru, A. Wave-Propagation and Scattering in Random-Media and Rough Surfaces. *Proc. IEEE* **1991**, *79*, 1359–1366. [CrossRef]
16. Ticconi, F.; Pulvirenti, L.; Pierdicca, N. Models for Scattering from Rough Surfaces. *Electromagn. Waves* **2011**. [CrossRef]
17. Hermansson, P.; Forssell, G. A Review of Models for Scattering from Rough Surfaces. Available online: https://scholar.google.com/scholar?cluster=9344336395593944459&hl=fr&as_sdt=0,5&sciodt=0,5 (accessed on 25 March 2020).
18. Ogilvy, J.A. Wave scattering from rough surfaces. *Rep. Prog. Phys.* **1987**, *50*, 1553–1608. [CrossRef]
19. Brekhovskikh, L.M.; Lysanov, Y.P. *Fundamentals of Ocean Acoustics*; Springer Science & Business Media: Berlin/Heidelberg, Germany, 2013; ISBN 978-3-662-07328-5.
20. Voronovich, A.G. *Wave Scattering from Rough Surfaces*; Springer Science & Business Media: Berlin/Heidelberg, Germany, 2013; ISBN 978-3-642-59936-1.
21. Thorsos, E.I. The validity of the Kirchhoff approximation for rough surface scattering using a Gaussian roughness spectrum. *J. Acoust. Soc. Am.* **1988**, *83*, 78–92. [CrossRef]
22. Khenchaf, A.; Daout, F.; Saillard, J. The two-scale model for random rough surface scattering. In Proceedings of the OCEANS 96 MTS/IEEE Conference Proceedings. The Coastal Ocean-Prospects for the 21st Century, Fort Lauderdale, FL, USA, 23–26 September 1996; IEEE: Piscataway, NY, USA, 1996; Volume 2, pp. 887–891.
23. Nayak, P.R. Random Process Model of Rough Surfaces. *J. Lubr. Technol.* **1971**, *93*, 398–407. [CrossRef]
24. Johnson, K.L.; Johnson, K.L. *Contact Mechanics*; Cambridge University Press: Cambridge, UK, 1987; ISBN 978-0-521-34796-9.
25. Adler, R.J.; Firmin, D.; Kendall, D.G. A non-gaussian model for random surfaces. *Philos. Trans. R. Soc. Lond. Ser. Math. Phys. Sci.* **1981**, *303*, 433–462. [CrossRef]
26. Pérez-Ràfols, F.; Almqvist, A. Generating randomly rough surfaces with given height probability distribution and power spectrum. *Tribol. Int.* **2019**, *131*, 591–604. [CrossRef]
27. Ogilvy, J.A. Computer simulation of acoustic wave scattering from rough surfaces. *J. Phys. Appl. Phys.* **1988**, *21*, 260–277. [CrossRef]
28. Civa Software Website. Available online: http://www.extende.com/ (accessed on 20 November 2020).
29. Mahaut, S.; Chatillon, S.; Darmon, M.; Leymarie, N.; Raillon, R.; Calmon, P. An overview of ultrasonic beam propagation and flaw scattering models in the civa software. *AIP Conf. Proc.* **2010**, *1211*, 2133–2140. [CrossRef]
30. Toullelan, G.; Raillon, R.; Chatillon, S.; Dorval, V.; Darmon, M.; Lonné, S. Results of the 2015 UT Modeling Benchmark Obtained with Models Implemented in Civa. *AIP Conf. Proc.* **2016**. [CrossRef]
31. Raillon-Picot, R.; Toullelan, G.; Darmon, M.; Calmon, P.; Lonné, S. Validation of CIVA ultrasonic simulation in canonical configurations. In Proceedings of the World Conference of Non Destructive Testing (WCNDT) 2012, Durban, South Africa, 16–20 April 2012.

32. Raillon, R.; Bey, S.; Dubois, A.; Mahaut, S.; Darmon, M. Results of the 2010 ut modeling benchmark obtained with civa: Responses of backwall and surface breaking notches. *AIP Conf. Proc.* **2011**, *1335*, 1777–1784. [CrossRef]
33. Raillon, R.; Bey, S.; Dubois, A.; Mahaut, S.; Darmon, M. Results of the 2009 ut modeling benchmark obtained with civa: Responses of notches, side-drilled holes and flat-bottom holes of various sizes. *AIP Conf. Proc.* **2010**, *1211*, 2157–2164. [CrossRef]
34. Gilbert, F.; Knopoff, L. Seismic scattering from topographic irregularities. *J. Geophys. Res. 1896–1977* **1960**, *65*, 3437–3444. [CrossRef]
35. Bass, F.G.; Fuks, I.M. *Wave Scattering from Statistically Rough Surfaces*; Pergamon Press: Oxford, UK, 1973.
36. Blakemore, M. Blakemore Scattering of acoustic waves by the rough surface of an elastic solid. *Ultrasonics* **1993**, *31*, 161–174. [CrossRef]
37. De Billy, M.; Quentin, G. Backscattering of acoustic waves by randomly rough surfaces of elastic solids immersed in water. *J. Acoust. Soc. Am.* **1982**, *72*, 591–601. [CrossRef]
38. Kuperman, W.A.; Schmidt, H. Self-consistent perturbation approach to rough surface scattering in stratified elastic media. *J. Acoust. Soc. Am.* **1989**, *86*, 1511–1522. [CrossRef]
39. Thorsos, E.I.; Jackson, D.R.; Williams, K.L. Modeling of subcritical penetration into sediments due to interface roughness. *J. Acoust. Soc. Am.* **1999**, *107*, 263–277. [CrossRef]
40. Jackson, D.R.; Ivakin, A.N. Scattering from elastic sea beds: First-order theory. *J. Acoust. Soc. Am.* **1998**, *103*, 336–345. [CrossRef]
41. Liu, J.Y.; Huang, C.F.; Hsueh, P.C. Acoustic plane-wave scattering from a rough surface over a random fluid medium. *Ocean Eng.* **2002**, *29*, 915–930. [CrossRef]
42. Liu, J.Y.; Tsai, S.H.; Wang, C.C.; Chu, C.R. Acoustic wave reflection from a rough seabed with a continuously varying sediment layer overlying an elastic basement. *J. Sound Vib.* **2004**, *275*, 739–755. [CrossRef]
43. Tang, D.; Jackson, D. A time-domain model for seafloor scattering. *J. Acoust. Soc. Am.* **2017**, *142*, 2968–2978. [CrossRef] [PubMed]
44. Bjorno, L. Scattering of plane acoustic waves at elastic particles with rough surfaces. *J. Acoust. Soc. Am.* **2015**, *137*, 2439. [CrossRef]
45. Neighbors, T.; Bradley, D. *Applied Underwater Acoustics: Leif Bjørnø*; Elsevier: Amsterdam, The Netherlands, 2017; ISBN 978-0-12-811247-2.
46. Faran, J.J. Sound Scattering by Solid Cylinders and Spheres. *J. Acoust. Soc. Am.* **1951**, *23*, 405–418. [CrossRef]
47. Flax, L.; Dragonette, L.R.; Überall, H. Theory of elastic resonance excitation by sound scattering. *J. Acoust. Soc. Am.* **1978**, *63*, 723–731. [CrossRef]
48. Masood, F.; Amin, U.; Fiaz, M.A.; Ashraf, M.A. On relating the perturbation theory and random cylinder generation to study scattered field. *Phys. Commun.* **2020**, *39*, 101003. [CrossRef]
49. Winebrenner, D.; Ishimaru, A. Investigation of a surface field phase perturbation technique for scattering from rough surfaces. *Radio Sci.* **1985**, *20*, 161–170. [CrossRef]
50. Winebrenner, D.P.; Ishimaru, A. Application of the phase-perturbation technique to randomly rough surfaces. *JOSA A* **1985**, *2*, 2285–2294. [CrossRef]
51. Broschat, S.L.; Tsang, L.; Ishimaru, A.; Thorsos, E.I. A Numerical Comparison of the Phase Perturbation Technique with the Classical Field Perturbation and Kirchhoff Approximations for Random Rough Surface Scattering. *J. Electromagn. Waves Appl.* **1988**, *2*, 85–102. [CrossRef]
52. Broschat, S.L.; Thorsos, E.I.; Ishimaru, A. The Phase Perturbation Technique vs. an Exact Numerical Method for Random Rough Surface Scattering. *J. Electromagn. Waves Appl.* **2012**. [CrossRef]
53. Zhang, X.D.; Wu, Z.S.; Wu, C.K. A phase-perturbation technique for light scattering from randomly dielectric rough surfaces. *Chin. Phys. Lett.* **1997**, *14*, 32–35.
54. Meecham, W.C. On the use of the Kirchhoff approximation for the solution of reflection problems. *Eng. Res. Inst. Dep. Phys. Mich. Univ.* **1955**, *5*, 323–334. [CrossRef]
55. Bouche, D.; Molinet, F.; Mittra, R. *Asymptotic Methods in Electromagnetics*; Springer: Berlin/Heidelberg, Germany, 1997.
56. Chungang, J.; Lixin, G.; Pengju, Y. Time-Domain Physical Optics Method for the Analysis of Wide-Band EM Scattering from Two-Dimensional Conducting Rough Surface. Available online: https://www.hindawi.com/journals/ijap/2013/584260/ (accessed on 21 September 2020).

57. Dorval, V.; Chatillon, S.; Lu, B.; Darmon, M.; Mahaut, S. A general Kirchhoff approximation for echo simulation in ultrasonic NDT. *AIP Conf. Proc.* **2012**, *1430*, 193–200. [CrossRef]
58. Darmon, M.; Leymarie, N.; Chatillon, S.; Mahaut, S. Modelling of Scattering of Ultrasounds by Flaws for NDT. In *Ultrasonic Wave Propagation in Non Homogeneous Media*; Springer: Berlin/Heidelberg, Germany, 2009; Volume 128, pp. 61–71.
59. Lu, B.; Darmon, M.; Potel, C.; Zernov, V. Models Comparison for the scattering of an acoustic wave on immersed targets. *J. Phys. Conf. Ser.* **2012**, *353*, 012009. [CrossRef]
60. Keller, J.B. Geometrical theory of diffraction. *J. Opt. Soc. Am.* **1962**, *52*, 116–130. [CrossRef]
61. Darmon, M.; Chatillon, S.; Mahaut, S.; Fradkin, L.; Gautesen, A. Simulation of disoriented flaws in a TOFD technique configuration using GTD approach. In Proceedings of the 34th Annual Review of Progress in Quantitative Nondestructive Evaluation, Chicago, IL, USA, 20–25 July 2008; Volume 975, pp. 155–162.
62. Chaffai, S.; Darmon, M.; Mahaut, S.; Menand, R. Simulations tools for TOFD inspection in Civa software. In Proceedings of the ICNDE 2007, Istanbul, Turkey, 15–20 April 2007.
63. Darmon, M.; Ferrand, A.; Dorval, V.; Chatillon, S.; Lonné, S. Recent Modelling Advances for Ultrasonic TOFD Inspections. In *AIP Conference Proceedings*; AIP Publishing: College Park, MD, USA, 2015; Volume 1650, pp. 1757–1765.
64. Chehade, S.; Kamta Djakou, A.; Darmon, M.; Lebeau, G. The spectral functions method for acoustic wave diffraction by a stress-free wedge: Theory and validation. *J. Comput. Phys.* **2019**, *377*, 200–218. [CrossRef]
65. Chehade, S.; Darmon, M.; Lebeau, G. 2D elastic plane-wave diffraction by a stress-free wedge of arbitrary angle. *J. Comput. Phys.* **2019**, *394*, 532. [CrossRef]
66. Ufimtsev, P.Y. *Fundamentals of the Physical Theory of Diffraction*; John Wiley & Sons: Hoboken, NJ, USA, 2007.
67. Lü, B.; Darmon, M.; Fradkin, L.; Potel, C. Numerical comparison of acoustic wedge models, with application to ultrasonic telemetry. *Ultrasonics* **2016**, *65*, 5–9. [CrossRef]
68. Zernov, V.; Fradkin, L.; Darmon, M. A refinement of the Kirchhoff approximation to the scattered elastic fields. *Ultrasonics* **2012**, *52*, 830–835. [CrossRef]
69. Darmon, M.; Dorval, V.; Kamta Djakou, A.; Fradkin, L.; Chatillon, S. A system model for ultrasonic NDT based on the Physical Theory of Diffraction (PTD). *Ultrasonics* **2016**, *64*, 115–127. [CrossRef] [PubMed]
70. Fradkin, L.J.; Darmon, M.; Chatillon, S.; Calmon, P. A semi-numerical model for near-critical angle scattering. *J. Acoust. Soc. Am.* **2016**, *139*, 141–150. [CrossRef] [PubMed]
71. Djakou, A.K.; Darmon, M.; Fradkin, L.; Potel, C. The Uniform geometrical Theory of Diffraction for elastodynamics: Plane wave scattering from a half-plane. *J. Acoust. Soc. Am.* **2015**, *138*, 3272–3281. [CrossRef] [PubMed]
72. Eckart, C. The Scattering of Sound from the Sea Surface. *J. Acoust. Soc. Am.* **1953**, *25*, 566–570. [CrossRef]
73. Imbert, C. *Visualisation Ultrasonore Rapide Sous Sodium Application Aux Reacteurs a Neutrons Rapides*; INSA de Lyon: Villeurbanne, France, 1997.
74. Wagner, R.J. Shadowing of Randomly Rough Surfaces. *J. Acoust. Soc. Am.* **1967**, *41*, 138–147. [CrossRef]
75. Zverev, V.A.; Slavinskii, M.M. A method for calculating the acoustic field near a rough surface. *Acoust. Phys.* **1997**, *43*, 56–60.
76. Chapman, D.M.F. An improved Kirchhoff formula for reflection loss at a rough ocean surface at low grazing angles. *J. Acoust. Soc. Am.* **1983**, *73*, 520–527. [CrossRef]
77. Shi, F.; Choi, W.; Lowe, M.J.S.; Skelton, E.A.; Craster, R.V. The validity of Kirchhoff theory for scattering of elastic waves from rough surfaces. *Proc. R. Soc. Math. Phys. Eng. Sci.* **2015**, *471*, 20140977. [CrossRef]
78. Haslinger, S.G. Elastic shear wave scattering by randomly rough surfaces. *J. Mech. Phys. Solids* **2020**, *137*, 103852. [CrossRef]
79. Haslinger, S.G.; Lowe, M.J.S.; Huthwaite, P.; Craster, R.V.; Shi, F. Appraising Kirchhoff approximation theory for the scattering of elastic shear waves by randomly rough defects. *J. Sound Vib.* **2019**, *460*, 114872. [CrossRef]
80. Opsal, J.L.; Visscher, W.M. Theory of elastic wave scattering: Applications of the method of optimal truncation. *J. Appl. Phys.* **1985**, *58*, 1102–1115. [CrossRef]
81. Becache, E.; Joly, P.; Tsogka, C. An analysis of new mixed finite elements for the approximation of wave propagation problems. *SIAM J. Numer. Anal.* **2000**, *37*, 1053–1084. [CrossRef]
82. Becache, E.; Joly, P.; Tsogka, C. Application of the fictitious domain method to 2D linear elastodynamic problems. *J. Comput. Acoust.* **2001**, *9*, 1175–1202.

83. Huang, R.; Schmerr, L.W., Jr.; Sedov, A.; Gray, T.A. Kirchhoff approximation revisited—Some new results for scattering in isotropic and anisotropic elastic solids. *Res. Nondestruct. Eval.* **2006**, *17*, 137–160. [CrossRef]
84. Darmon, M. *Validity of the Kirchhoff Approximation for Small Flaws*; CEA/DISC/LSMA: Saclay, France, 2014.
85. Zhang, J.; Drinkwater, B.W.; Wilcox, P.D. Longitudinal Wave Scattering from Rough Crack-Like Defects. *IEEE Trans. Ultrason. Ferroelectr. Freq. Control* **2011**, *58*, 2171–2180. [CrossRef]
86. Zernov, V.; Fradkin, L.; Gautesen, A.; Darmon, M.; Calmon, P. Wedge diffraction of a critically incident Gaussian beam. *Wave Motion* **2013**, *50*, 708–722. [CrossRef]
87. Zernov, V.; Gautesen, A.; Fradkin, L.J.; Darmon, M. Aspects of diffraction of a creeping wave by a back-wall crack. *J. Phys. Conf. Ser.* **2012**, *353*, 012017. [CrossRef]
88. Mahaut, S.; Huet, G.; Darmon, M. Modeling of Corner Echo in UT Inspection Combining Bulk and Head Waves Effect. In Proceedings of the 35th Annual Review of Progress in Quantitative Nondestructive Evaluation, 35th Annual Review of Progress in Quantitative Nondestructive Evaluation, Chicago, IL, USA, 20–25 July 2008; Volume 1096, pp. 73–80.
89. Fradkin, L.J.; Djakou, A.K.; Prior, C.; Darmon, M.; Chatillon, S.; Calmon, P.-F. The Alternative Kirchhoff Approximation in Elastodynamics with Applications in Ultrasonic Nondestructive Testing. *ANZIAM J.* **2020**, 1–17. [CrossRef]
90. Ferrand, A.; Darmon, M.; Chatillon, S.; Deschamps, M. Modeling of ray paths of head waves on irregular interfaces in TOFD inspection for NDE. *Ultrasonics* **2014**, *54*, 1851–1860. [CrossRef]
91. Bjorno, L.; Sun, S. Use of the Kirchhoff Approximation in Scattering from Elastic, Rough Surfaces. *Acoust. Phys.* **1995**, *41*, 637–648.
92. Dacol, D.K. The Kirchhoff approximation for acoustic scattering from a rough fluid–elastic solid interface. *J. Acoust. Soc. Am.* **1998**, *88*, 978. [CrossRef]
93. Gavrilov, A.L.; Dunin, S.Z.; Maskomov, G.A. Scattering of scalar fields by hard and soft rough surfaces. Angular distribution of intensity. *Akust. Zhurnal* **1992**, *38*, 861–873.
94. Ploix, M.-A.; Kauffmann, P.; Chaix, J.-F.; Lillamand, I.; Baqué, F.; Corneloup, G. Acoustical properties of an immersed corner-cube retroreflector alone and behind screen for ultrasonic telemetry applications. *Ultrasonics* **2020**, *106*. [CrossRef] [PubMed]
95. Darmon, M.; Chatillon, S. Main Features of a Complete Ultrasonic Measurement Model: Formal Aspects of Modeling of Both Transducers Radiation and Ultrasonic Flaws Responses. *Open J. Acoust.* **2013**, *3*, 36873. [CrossRef]
96. Mahaut, S.; Darmon, M.; Chatillon, S.; Jenson, F.; Calmon, P. Recent advances and current trends of ultrasonic modelling in CIVA. *Insight* **2009**, *51*, 78–81. [CrossRef]
97. Isakson, M.J.; Chotiros, N.P. Finite Element Modeling of Acoustic Scattering from Fluid and Elastic Rough Interfaces. *IEEE J. Ocean. Eng.* **2015**, *40*, 475–484. [CrossRef]
98. Botten, L.C.; Cadilhac, M.; Derrick, G.H.; Maystre, D.; McPhedran, R.C.; Nevière, M.; Vincent, P. *Electromagnetic Theory of Gratings*; Springer Science & Business Media: Berlin/Heidelberg, Germany, 2013; Volume 22.
99. Richards, E.L.; Song, H.C.; Hodgkiss, W.S. Acoustic scattering comparison of Kirchhoff approximation to Rayleigh-Fourier method for sinusoidal surface waves at low grazing angles. *J. Acoust. Soc. Am.* **2018**, *144*, 1269–1278. [CrossRef]
100. Williams, K.L.; Stroud, J.S.; Marston, P.L. High-frequency forward scattering from Gaussian spectrum, pressure release, corrugated surfaces. I. Catastrophe theory modeling. *J. Acoust. Soc. Am.* **1994**, *96*, 1687–1702. [CrossRef]
101. Kur'yanov, B.F. The scattering of sound at a rough surface with two types of irregularity. *Sov. Phys. Acoust.* **1963**, *8*, 252–257.
102. McDaniel, S.T.; Gorman, A.D. An examination of the composite-roughness scattering model. *J. Acoust. Soc. Am.* **1983**, *73*, 1476–1486. [CrossRef]
103. Bachmann, W. A theoretical model for the backscattering strength of a composite-roughness sea surface. *J. Acoust. Soc. Am.* **2005**, *54*, 712. [CrossRef]
104. Lemaire, D.; Sobieski, P.; Craeye, C.; Guissard, A. Two-scale models for rough surface scattering: Comparison between the boundary perturbation method and the integral equation method. *Radio Sci.* **2002**, *37*, 1–16. [CrossRef]

105. Novarini, J.C.; Caruthers, J.W. The partition wavenumber in acoustic backscattering from a two-scale rough surface described by a power-law spectrum. *IEEE J. Ocean. Eng.* **1994**, *19*, 200–207. [CrossRef]
106. Jones, A.D.; Sendt, J.S.; Duncan, A.J.; Zhang, Y.; Clarke, P.A.B. Modelling Acoustic Reflection Loss at the Ocean Surface—An Australian Study. Available online: /paper/Modelling-Acoustic-Reflection-Loss-at-the-Ocean-an-Jones-Sendt/697e50ae9aae2ed887353067e825c9b31bad19bc (accessed on 2 June 2020).
107. Voronovich, A.G. Small slope approximation in wave scattering theory for rough surfaces. *Zhurnal Eksp. Teor. Fiz.* **1985**, *89*, 116–125.
108. Voronovich, A. Small-slope approximation for electromagnetic wave scattering at a rough interface of two dielectric half-spaces. *Waves Random Media* **1994**, *4*, 337–367. [CrossRef]
109. Voronovich, A.G. Non-local small-slope approximation for wave scattering from rough surfaces. *Waves Random Media* **1996**, *6*, 151–167. [CrossRef]
110. McDaniel, S.T. A small-slope theory of rough surface scattering. *J. Acoust. Soc. Am.* **1994**, *95*, 1858–1864. [CrossRef]
111. Thorsos, E.I.; Broschat, S.L. An investigation of the small slope approximation for scattering from rough surfaces. Part I. Theory. *J. Acoust. Soc. Am.* **1995**, *97*, 2082–2093. [CrossRef]
112. Broschat, S.L.; Thorsos, E.I. An investigation of the small slope approximation for scattering from rough surfaces. Part II. Numerical studies. *J. Acoust. Soc. Am.* **1997**, *101*, 2615–2625. [CrossRef]
113. Salin, M.B.; Dosaev, A.S.; Konkov, A.I.; Salin, B.M. Numerical simulation of Bragg scattering of sound by surface roughness for different values of the Rayleigh parameter. *Acoust. Phys.* **2014**, *60*, 442–454. [CrossRef]
114. Yang, T.; Broschat, S.L. Acoustic scattering from a fluid–elastic-solid interface using the small slope approximation. *J. Acoust. Soc. Am.* **1994**, *96*, 1796–1804. [CrossRef]
115. Berman, D.H. Simulations of rough interface scattering. *J. Acoust. Soc. Am.* **1991**, *89*, 623–636. [CrossRef]
116. Jackson, D.; Olson, D.R. The small-slope approximation for layered, fluid seafloors. *J. Acoust. Soc. Am.* **2020**, *147*, 56–73. [CrossRef] [PubMed]
117. Jackson, D. The small-slope approximation for layered seabeds. *Proc. Meet. Acoust.* **2013**, *19*, 070001. [CrossRef]
118. Berrouk, A.; Dusseaux, R.; Afifi, S. Electromagnetic Wave Scattering from Rough Layered Interfaces: Analysis with the Small Perturbation Method and the Small Slope Approximation. *Prog. Electromagn. Res.* **2014**, *57*, 177–190. [CrossRef]
119. Afifi, S.; Dusséaux, R.; Berrouk, A. Electromagnetic Scattering from 3D Layered Structures with Randomly Rough Interfaces: Analysis With the Small Perturbation Method and the Small Slope Approximation. *IEEE Trans. Antennas Propag.* **2014**, *62*, 5200–5208. [CrossRef]
120. Dusséaux, R.; Afifi, S.; Dechambre, M. Scattering properties of a stratified air/snow/sea ice medium. Small slope approximation. *Comptes Rendus Phys.* **2016**, *17*, 995–1002. [CrossRef]
121. Collins, M.D. A split-step Padé solution for the parabolic equation method. *J. Acoust. Soc. Am.* **1993**, *93*, 1736–1742. [CrossRef]
122. Gao, Y.; Shao, Q.; Yan, B.; Li, Q.; Guo, S. Parabolic Equation Modeling of Electromagnetic Wave Propagation over Rough Sea Surfaces. *Sensors* **2019**, *19*, 1252. [CrossRef] [PubMed]
123. Thorsos, E.I. Rough surface scattering using the parabolic wave equation. *J. Acoust. Soc. Am.* **1987**, *82*, S103. [CrossRef]
124. Ramsurf Website. Available online: https://github.com/quiet-oceans/ramsurf (accessed on 17 October 2019).
125. Modeling Reflection Loss of the Gaussian Rough Ocean Surface—IEEE Conference Publication. Available online: https://ieeexplore.ieee.org/document/8559454 (accessed on 11 October 2019).
126. Jones, A.D.; Duncan, A.J.; Maggi, A.L.; Bartel, D.W.; Zinoviev, A. A Detailed Comparison between a Small-Slope Model of Acoustical Scattering from a Rough Sea Surface and Stochastic Modeling of the Coherent Surface Loss. *IEEE J. Ocean. Eng.* **2016**, *41*, 689–708. [CrossRef]
127. Spivack, M.; Rath Spivack, O. Rough Surface Scattering via Two-Way Parabolic Integral Equation. *Prog. Electromagn. Res.* **2017**, *56*, 81–90. [CrossRef]
128. Huet, G.; Darmon, M.; Lhemery, A.; Mahaut, S. Modelling of Corner Echo Ultrasonic Inspection with Bulk and Creeping Waves. In *5th Meeting of the Anglo-French-Research-Group*; Léger, A., Deschamps, M., Eds.; Springer: Berlin/Heidelberg, Germany, 2009; Volume 128, pp. 217–226.

129. Ferrand, A. Développement de Modèles Asymptotiques en Contrôle Non Destructif (CND) par Ultrasons: Interaction des Ondes Elastiques Avec des Irrégularités Géométriques et Prise en Compte des Ondes de Tête. Ph.D. Thesis, Université Sciences et Technologies, Bordeaux, France, 2014.
130. Zhou, H.; Chen, X. Ray path of head waves with irregular interfaces. *Appl. Geophys.* **2010**, *7*, 66–73. [CrossRef]
131. Olson, D.R. Numerical investigation of the two-scale model for rough surface scattering. *J. Acoust. Soc. Am.* **2019**, *145*, 1770. [CrossRef]
132. Rath Spivack, O.; Spivack, M. Efficient boundary integral solution for acoustic wave scattering by irregular surfaces. *Eng. Anal. Bound. Elem.* **2017**, *83*, 275–280. [CrossRef]
133. Li, C.; Campbell, B.K.; Liu, Y.; Yue, D.K.P. A fast multi-layer boundary element method for direct numerical simulation of sound propagation in shallow water environments. *J. Comput. Phys.* **2019**, *392*, 694–712. [CrossRef]
134. Lee, S.; Pegues, J.W.; Shamsaei, N. Fatigue behavior and modeling for additive manufactured 304L stainless steel: The effect of surface roughness. *Int. J. Fatigue* **2020**, *141*, 105856. [CrossRef]
135. Knezović, N.; Topić, A. Wire and Arc Additive Manufacturing (WAAM)—A New Advance in Manufacturing. In *Lecture Notes in Networks and Systems*; Springer International Publishing: New York, NY, USA, 2019; pp. 65–71; ISBN 978-3-319-90892-2.
136. Gu, D.D.; Meiners, W.; Wissenbach, K.; Poprawe, R. Laser additive manufacturing of metallic components: Materials, processes and mechanisms. *Int. Mater. Rev.* **2012**, *57*, 133–164. [CrossRef]
137. Gorsse, S.; Hutchinson, C.; Gouné, M.; Banerjee, R. Additive manufacturing of metals: A brief review of the characteristic microstructures and properties of steels, Ti-6Al-4V and high-entropy alloys. *Sci. Technol. Adv. Mater.* **2017**, *18*, 584–610. [CrossRef]
138. Tian, H.; Lu, Z.; Li, F.; Chen, S. *Predictive Modeling of Surface Roughness Based on Response Surface Methodology after WAAM*; Atlantis Press: Paris, France, 2019; pp. 47–50.
139. NUTALN. Finition de Surface de Pièces Produites par Fabrication. Additive. Available online: https://www.techniques-ingenieur.fr/base-documentaire/42687210-chaine-de-valeur-et-mise-en-uvre/download/bm7960/finition-de-surface-de-pieces-produites-par-fabrication-additive.html (accessed on 20 October 2020).
140. Gharbi, M.; Peyre, P.; Gorny, C.; CARIN, M.; Morville, S.; LE MASSON, P.; CARRON, D.; Fabbro, R. Influence of various process conditions on surface finishes induced by the direct metal deposition laser technique on a Ti-6Al-4V alloy. *J. Mater. Process. Technol.* **2012**, *213*, 791–800. [CrossRef]
141. Xiong, J.; Li, Y.-J.; Yin, Z.-Q.; Chen, H. Determination of Surface Roughness in Wire and Arc Additive Manufacturing Based on Laser Vision Sensing. *Chin. J. Mech. Eng.* **2018**, *31*, 74. [CrossRef]
142. Sanviemvongsak, T.; Monceau, D.; Macquaire, B. High temperature oxidation of IN 718 manufactured by laser beam melting and electron beam melting: Effect of surface topography. *Corros. Sci.* **2018**, *141*, 127–145. [CrossRef]
143. Taheri, H.; Koester, L.; Bigelow, T.; Bond, L.J. Finite element simulation and experimental verification of ultrasonic non-destructive inspection of defects in additively manufactured materials. *AIP Conf. Proc.* **2018**, *1949*, 020011. [CrossRef]
144. Van Pamel, A.; Nagy, P.B.; Lowe, M.J.S. On the dimensionality of elastic wave scattering within heterogeneous media. *J. Acoust. Soc. Am.* **2016**, *140*, 4360–4366. [CrossRef] [PubMed]
145. Ryzy, M.; Grabec, T.; Sedlák, P.; Veres, I.A. Influence of grain morphology on ultrasonic wave attenuation in polycrystalline media with statistically equiaxed grains. *J. Acoust. Soc. Am.* **2018**, *143*, 219–229. [CrossRef]
146. Bai, X.; Tie, B.; Schmitt, J.-H.; Aubry, D. Comparison of ultrasonic attenuation within two- and three-dimensional polycrystalline media. *Ultrasonics* **2020**, *100*, 105980. [CrossRef]
147. OUDAA, M.; Lhuillier, P.-E.; Guy, P.; Leclere, Q. Finite element modeling of ultrasonic attenuation within polycrystalline materials in two and three dimensions. In Proceedings of the 2019 International Congress on Ultrasonics, Bruges, Belgium, 3–6 September 2019.
148. Komatitsch, D.; Tromp, J. Introduction to the spectral element method for three-dimensional seismic wave propagation. *Geophys. J. Int.* **1999**, *139*, 806–822. [CrossRef]
149. Casadei, F.; Gabellini, E.; Fotia, G.; Maggio, F.; Quarteroni, A. A mortar spectral/finite element method for complex 2D and 3D elastodynamic problems. *Comput. Methods Appl. Mech. Eng.* **2002**, *191*, 5119–5148. [CrossRef]

150. Imperiale, A.; Chatillon, S.; Darmon, M.; Leymarie, N.; Demaldent, E. UT simulation using a fully automated 3D hybrid model: Application to planar backwall breaking defects inspection. *AIP Conf. Proc.* **2018**, *1949*, 050004. [CrossRef]
151. Yu, T.; Chaix, J.-F.; Audibert, L.; Komatitsch, D.; Garnier, V.; Henault, J.-M. Simulations of ultrasonic wave propagation in concrete based on a two-dimensional numerical model validated analytically and experimentally. *Ultrasonics* **2019**, *92*, 21–34. [CrossRef]
152. Hériveaux, Y.; Nguyen, V.-H.; Haïat, G. Reflection of an ultrasonic wave on the bone-implant interface: A numerical study of the effect of the multiscale roughness. *J. Acoust. Soc. Am.* **2018**, *144*, 488–499. [CrossRef] [PubMed]

MDPI
St. Alban-Anlage 66
4052 Basel
Switzerland
Tel. +41 61 683 77 34
Fax +41 61 302 89 18
www.mdpi.com

Applied Sciences Editorial Office
E-mail: applsci@mdpi.com
www.mdpi.com/journal/applsci

www.ingramcontent.com/pod-product-compliance
Lightning Source LLC
LaVergne TN
LVHW070622170726
843515LV00004B/153